Modern Deep Learning Design and Application Development

Versatile Tools to Solve Deep Learning Problems

Andre Ye

Apress®

Modern Deep Learning Design and Application Development: Versatile Tools to Solve Deep Learning Problems

Andre Ye
Redmond, WA, USA

ISBN-13 (pbk): 978-1-4842-7412-5   ISBN-13 (electronic): 978-1-4842-7413-2
https://doi.org/10.1007/978-1-4842-7413-2

Managing Director, Apress Media LLC: Welmoed Spahr
Acquisitions Editor: Celestin Suresh John
Development Editor: James Markham
Coordinating Editor: Aditee Mirashi

Cover designed by eStudioCalamar

Cover image designed by Freepik (www.freepik.com)

Distributed to the book trade worldwide by Springer Science+Business Media New York, 1 New York Plaza, Suite 4600, New York, NY 10004-1562, USA. Phone 1-800-SPRINGER, fax (201) 348-4505, e-mail orders-ny@springer-sbm.com, or visit www.springeronline.com. Apress Media, LLC is a California LLC and the sole member (owner) is Springer Science + Business Media Finance Inc (SSBM Finance Inc). SSBM Finance Inc is a **Delaware** corporation.

For information on translations, please e-mail booktranslations@springernature.com; for reprint, paperback, or audio rights, please e-mail bookpermissions@springernature.com.

Apress titles may be purchased in bulk for academic, corporate, or promotional use. eBook versions and licenses are also available for most titles. For more information, reference our Print and eBook Bulk Sales web page at http://www.apress.com/bulk-sales.

Any source code or other supplementary material referenced by the author in this book is available to readers on GitHub via the book's product page, located at www.apress.com/978-1-4842-7412-5. For more detailed information, please visit http://www.apress.com/source-code.

Printed on acid-free paper

For my mother Fang and my father Quinn

Table of Contents

About the Author

Andre Ye is a deep learning researcher and writer working toward making deep learning more accessible, understandable, and responsible through technical communication. He is also a cofounder at Critiq, a machine learning platform facilitating greater efficiency in the peer-review process. In his spare time, Andre enjoys keeping up with current deep learning research, reading up on history and philosophy, and playing the piano.

About the Technical Reviewer

Manohar Swamynathan is a data science practitioner and an avid programmer, with over 14+ years of experience in various data science-related areas that include data warehousing, Business Intelligence (BI), analytical tool development, ad hoc analysis, predictive modeling, data science product development, consulting, formulating strategies, and executing analytics programs. He's had a career covering the life cycle of data across different domains, such as US mortgage banking, retail/ecommerce, insurance, and industrial IoT. He has a bachelor's degree with a specialization in physics, mathematics, and computers and a master's degree in project management. He's currently living in Bengaluru, the Silicon Valley of India.

Acknowledgments

This book could not have been developed and published without the direct and indirect help of so many. I want to express my thanks to Celestin Suresh John, who first made me aware of this opportunity; Aditee Mirashi, the coordinating editor who was always there to assist me and provide help to my countless inquiries; Jim Markham, the development editor, and Manohar Swamynathan, the technical reviewer, both of whom provided helping suggestions and comments in revision; and the editing team broadly that made writing this book such a wonderful experience. I also would like to thank the authors and coauthors of the over 20 deep learning papers used in this book as case studies for their cooperation and help.

Throughout writing the book, my family has always been there for me – my father Quinn, my mother Fang, and my brother Aiden. It was my father's early work in neural networks and support that spurred much of my initial interest in deep learning, and my family has been incredibly welcoming of conversation and discussion on the subject and for emotional support more generally. I also want to appreciate Andy Wang, who has been my go-to person for deep learning ideas, chatting, and project collaboration, and my friends – Krishna, Mark, Ayush, and Vivek – for their unwavering support, friendship, and encouragement.

Lastly, I would like to thank the Robinson Center – a truly wonderful accelerator for intellectual development – and the amazing faculty and staff there that exposed me to new and profound ideas and perspectives of thinking in my formative years. Their involvement, I believe, has made me an improved thinker and communicator of knowledge, technical and otherwise.

Introduction

Deep learning is almost universally regarded as falling under the field of computer science, with some intersections in statistics and mathematics broadly, like most computer science disciplines. Thus, upon first impression, it may be perceived to be a strict, exacting science – a rigid process of adjusting architectures and methods in a rote fashion to squeeze out a few more percentage points of accuracy on some dataset. While this is arguably a necessary component and period of deep learning's trajectory of growth and development, if embraced too wholeheartedly, it becomes damaging. Such a perspective is lethal toward innovation in deep learning, for many – and, possibly, most – of the major advances in deep learning were and are not derived through a strict, rigorous scientific process of adjusting but rather through wild and free ideas.

Deep learning, as will be further explicated in this book, has always been somewhat of an experiment-first explanation-after field. That is, the success of a method is almost always explained in fuller detail after the method itself has been empirically observed to be successful. The method itself usually emerges from a hunch or intuition developed through a long series of experimentation and revision rather than a carefully constructed system in which a strong and rational explanation for a method's success precedes its empirical success. This is not, of course, to suggest that to make strides in deep learning one must be *messy*; it is only to propose that deep learning is a study in a stage guided primarily by the spirit of freedom in thought and experimentation.

Part of a different but seemingly more popular perspective of deep learning as rigid and more or less a continuation of the "find the best model" paradigm within machine learning, undoubtedly, arises from an immense focus on validation metrics that was spurred in the relatively early development of more advanced and powerful deep learning architectures and methods in the early to middle 2010s and the continuation of this spirit in popular deep learning competitions and competition platforms.

Beyond this phase, though, deep learning is often presented in a way that encourages this problematic perspective. Books, courses, online articles, and other mediums explain deep learning through a series of concrete examples, packaging their primary value as lying within the implementation of said examples. Concrete examples are important and come at little cost in a digital-based field, which is likely why they have become so prized in the computer science context.

The trouble here is that the focus is often too narrowly placed upon specific examples rather than broader concepts. It is very much reflective of the "give a man a fish and you feed him for a day, teach him how to fish and you feed him for his lifetime" aphorism. Rather than feeding single fish to readers, teaching motion and freedom – the act of fishing for new ideas and applications – must be a key component in the presentation of deep learning. Examples and implementations that are shown should be anchor points that guide broader conceptual understandings rather than being the sole occupants of the reader's understanding.

Think of this in relation to the overfitting paradigm introduced early to machine learning practitioners. If a simple curve fitting model places too high a weight on passing through each point in its dataset, it models the data perfectly but fails to model the underlying phenomena – the *concept* – that the data represents. This is overfitting. The model is thus prone to difficulty generalizing the ideas that the data represents, failing to be useful in new applications and contexts. Many of the measures used in classical machine learning as well as deep learning seek to maximize the model's performance with respect to modeling the *concepts* underlying the data rather than the data in and of itself. Likewise, the presentation of deep learning will inevitably be heavily tied to specific examples and developments, but should not consist *just* of these specific examples, for someone who has learned deep learning this way will likely have difficulty using deep learning in a context outside the narrow range that they learned those examples in. Moreover, they will find it challenging to innovate new ideas and approaches – a skill incredibly valuable in such a quickly evolving field.

The goal of this book is to take a step toward this concept-prioritizing vision of deep learning communication, to communicate the *ideas* of deep learning *design*, rather than to develop the reader's knowledge of deep learning in a manner consisting only of specific architectures and methods with no understanding of how to generalize and find innovative freedom within deep learning.

Make no mistake – this goal does not mean that this book is lacking in examples or implementation. The first chapter, "A Deep Dive into Keras," is dedicated entirely toward developing a deep understanding of Keras, the deep learning library we will use to realize concepts as implementations and applications. The remainder of the book is organized generally in an intuition/theory-first, implementation-second bipartite structure; the concept is introduced first with examples, analogy, and necessary instruments of theory and intuition, followed by a walk-through of code implementations and variations. Additionally, each of the six chapters following the first contains three case studies, each of which covers a deep learning paper from recent deep learning research in a perspective relevant to the chapter's topic, some of which contain code walk-throughs. The objective of the case studies is multifold: to explore the diversity of topics explored in current deep learning research, to provide empirical evidence on the success of certain methods, and to provide examples of deep learning innovations that push the boundaries and correspondingly advance the field's state.

The six chapters following the first each cover an important set of concepts in modern deep learning. It is important to note that these chapters assume a basic knowledge of neural network operation (feed-forward, gradient descent, standard, and convolutional layers) and general machine learning principles. The chapters successively build upon these foundations to explore transfer learning and self-supervised learning in Chapter 2, "Pretraining Strategies and Transfer Learning"; autoencoders in Chapter 3, "The Versatility of Autoencoders"; model compression in Chapter 4, "Model Compression for Practical Deployment"; Bayesian optimization and Neural Architecture Search in Chapter 5, "Automating Model Design with Meta-optimization"; modern architecture design in Chapter 6, "Successful Neural Network Architecture Design"; and problem-solving in Chapter 7, "Reframing Difficult Deep Learning Problems." Readers, especially those familiar with other deep learning resources, will find that these chapters contain many undiscussed topics in other books and tutorials, including self-supervised learning; model compression methods like pruning, quantization, and weight clustering; and Neural Architecture Search methods. Broadly, the book attempts a novel approach at communicating both core and boundary-pushing aspects of deep learning in a light that emphasizes intuition, variation, and generalization across concepts and methods.

The hope is that this book not only gives the reader a strong, deep understanding of the topics explored in each chapter but also intellectual freedom to propose and pursue new deep learning ideas addressing the great expanse of difficult problems we face today and will face tomorrow.

## CHAPTER 1

A Deep Dive into Keras

Imagination is more important than knowledge. For knowledge is limited, whereas imagination embraces the entire world.

—Albert Einstein[1]

One of the key goals of this book is to explore concepts in deep learning as an analytical art – the computing environment is your canvas, and deep learning design is your brush. A strong grasp of analytical creativity and imagination is integral to understanding deep learning at an instinctive, visceral level. This book aims to provide you with that understanding. As the world of deep learning constantly changes, those that attach their knowledge of deep learning to only the tools of today will find it difficult to work with deep learning when new syntaxes and libraries are introduced. An intuitive sense of playfulness and intellectual freedom and curiosity, though, can get us closer to working with deep learning closely, personally, and successfully beyond an interdependence on any one specific model architecture or framework.

Of course, interdependence is a lofty ideal; yet, in response to Einstein's famous quote, imagination and creativity must be grounded in certain concrete sets of knowledge to exist at all. Being able to explore and understand the analytical art of deep learning first requires setting up the framework for implementation, much like laying the foundation for a house. The knowledge – from concepts to code – will allow us to explore more abstracted and complex ideas in deep learning in later chapters.

---

[1] https://store.doverpublications.com/0486470105.html.

© Andre Ye 2022
A. Ye, *Modern Deep Learning Design and Application Development*,
https://doi.org/10.1007/978-1-4842-7413-2_1

This chapter will take a deep dive into Keras, the chosen framework for understanding how to build and implement the concepts we'll develop. The content covered in this chapter is not intended to be a substitute for easily accessible online syntax documentation – Keras and TensorFlow API docs are the best for those purposes. Rather, it is meant both as a quick introduction to the key building blocks of Keras to build more complex structures we'll work with later and as a concept reference guide that may be valuable to refer back to later. Beyond covering the syntax of Keras – as most online code tutorials or API docs do – we'll dive into frameworks to understand tips, tricks, and concepts to ensure efficient and masterful usage of Keras and the process of thinking.

Why Keras?

In Greek, Keras (κέρας) means "horn," in reference to the *Odyssey*. It was initially developed for project ONEIROS (Open-ended Neuro-Electronic Intelligent Robot Operating System) but has since become exceedingly popular in the modern deep learning community for a wide scope of applications and contexts.

There are many deep learning frameworks available, but Keras is particularly remarkable in its simplicity and versatility. Keras brands itself as "deep learning for humans," but its user-friendliness speaks for itself. Keras' design minimizes the time and energy required to bridge idea and implementation.

A 2019 survey asked top-scoring teams in Kaggle competitions about which machine learning frameworks they used and found that Keras ranked first.[2] Francis Chollet, the creator of Keras, wrote in a Quora post about his thoughts on why this was:

> *Developing good models requires iterating many times on your initial ideas, up until the deadline; you can always improve your models further... Keras was designed as a way to quickly prototype many different models, with a focus on reducing as much as possible the time it takes to go from having an idea to experimental results.*

—Francis Chollet[3]

---

[2] https://keras.io/why_keras/.

[3] www.quora.com/Why-has-Keras-been-so-successful-lately-at-Kaggle-competitions.

Keras' speed of development and versatility also makes it a popular choice for companies like Netflix and Instacart, researchers, and scientific organizations like CERN and NASA. As of early 2021, Keras has over 400,000 users.[4]

This is not a book fundamentally about Keras – it's a book about modern deep learning design. An integral part of design, though, as discussed, is implementation. Keras will be the valuable medium through which we can frame and realize the ideas we discuss. With Keras, we are freer to explore the possibilities of deep learning without being tied to arduous and lengthy implementations.

Installing and Importing Keras

Keras can be installed by itself via `pip` as `pip install keras` and imported as `import keras`. This (and all future installments) can be installed in the command prompt interface directly or in Jupyter Notebook by adding an exclamation mark (!) before the command. However, Keras has become part of TensorFlow, a deep learning framework by Google that allows for more low-level control. Since this merger, Keras users can take advantage of both Keras' user-friendliness and TensorFlow's lower-level customization abilities.

Since TensorFlow includes Keras, it's redundant to install Keras separately. Thus, often only TensorFlow is installed, and Keras is imported from TensorFlow (Listing 1-1).

Listing 1-1. Installing and importing TensorFlow and Keras

```
!pip install tensorflow      # only install TF
import tensorflow as tf      # import TF
from tensorflow import keras # import keras from TF
```

If you do not already have a workspace, consider the following to get started:

- *Jupyter Notebook*: Jupyter Notebook is especially popular for data science because its cell-based input-output structure allows for rapid experimentation and clean organization. Installing via Anaconda automatically installs important data science libraries.

---

[4] https://keras.io/why_keras/.

- *Kaggle*: Kaggle is an online data science platform that hosts competitions, datasets, data science forums, and code notebooks. Kaggle notebooks support both Jupyter Notebooks and Python or R scripts. Although Kaggle cannot handle highly computation or memory intensive operations, as it is web-based, it provides easy-to-access GPU and TPU. Moreover, Kaggle allows you to easily run notebooks for several hours on their own servers, which is a simple alternative to setting up virtual machines or other computational workflows. As of 2021, Kaggle usually puts a 30- to 42-hour weekly cap on GPU usage (varying by week) and a 30-hour weekly cap on TPU usage.

- *Google Colab*: Google Colab also allows for GPU and TPU usage, which is technically unlimited but can be cut off when the available computational resources cannot handle the demand and you've already been consuming the resources for a significant period of time. Google Colab is good for sharing and for quick experimentation on high-powered computation.

The Simple Keras Workflow

When building any model with Keras, there are generally three steps in the deep learning workflow: define architecture, compile, and fit and analyze (Figure 1-1). The simplest to the most complex models follow this workflow. In this section, we'll review each of the steps in this workflow in detail.

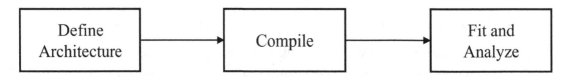

Figure 1-1. *The input-body-output model of a neural network*

Step 1: Define Architecture

There are three key parts of a neural network's architecture (Figure 1-2).

- *Input*: This part defines and receives the format of the input data that the neural network will use for the basis of prediction.

- *Body*: This part performs a series of operations on the input to transform it into an output. The body is composed of the layers of the neural network.

- *Output*: This part defines the format of the neural network's output result.

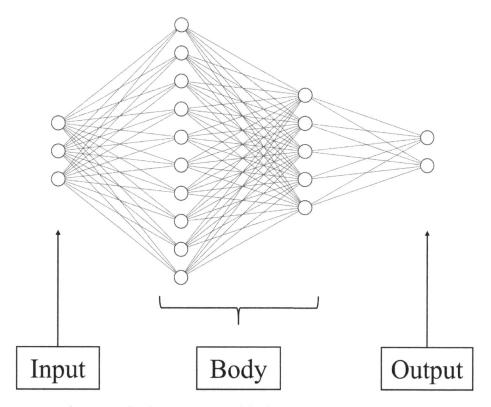

Figure 1-2. *The input-body-output model of a neural network*

Note   In this section, we will use the Sequential Keras API to illustrate the Keras workflow. The Sequential model allows the user to build models by stacking on layers linearly or sequentially. It suffices for simpler architectures, but more complex designs require the Functional API, which will be discussed later.

Let's consider a very simple dataset to define the architecture of a neural network (Table 1-1). This sample dataset contains three binary features – A, B, and C – and two binary labels. Label 1 is 1 only if any of the features is 1, and Label 2 is 1 only if at least two of the features have value 1.

Table 1-1. *Sample dummy dataset with three features and two labels*

A	B	C	Label 1	Label 2
0	0	0	0	0
1	0	0	1	0
0	1	0	1	0
0	1	1	1	1
1	1	0	1	1
1	1	1	1	1

This dataset has three features, which means that the input is three-dimensional. Let's begin by importing Keras, the Sequential model, and the Input layer; we can create a model and define the input (Listing 1-2).

Listing 1-2. Building the input component of a Sequential model

```
import keras
from keras.models import Sequential
from keras.layers import Input

model = Sequential() # initializes sequential model
model.add(Input((3,)))
```

Note   If the model were to handle images, the input shape may look something like Input((128,128,3)) for a 128x128 pixel colored image, in which each pixel holds three values corresponding to a color. The specifics of the shape of your image, like the difference between the shapes (3,) and (3,1), depend on how you handled your data. You can always use x_data[0].shape to find the specific shape of your data. You'll see more examples of this in later chapters that demonstrate image-based deep learning methods.

Now that we've specified the shape of our input, we can begin adding layers to form the body of the neural network. This simple neural network will have only Dense layers, which are regular densely connected neural network layers (Listing 1-3). Generally, two parameters are provided when using Dense layers – the number of units (neurons, nodes) in the layer and the activation function.

Listing 1-3. Building the body component of a Sequential model

```
from keras.layers import Dense

# layer with 10 nodes and ReLU activation
model.add(Dense(10, activation='relu'))

# layer with 5 nodes and ReLU activation
model.add(Dense(5, activation='relu'))
```

Note   The Dense layer is limited in its range of activation functions. To use advanced activation functions, do not provide an activation function in the standard densely connected layer (it will not apply one by default) and follow it with an Activation class layer, like model.add(keras.layers.LeakyReLU()).

These two layers comprise the body of the neural network – they do the heavy lifting of learning relationships to understand the mapping between the input and the output. In this case, the data is relatively simple, so this small body will suffice, but more difficult problems will require larger and more complex bodies. We'll become familiar with the design of more complex neural network architectures in the following chapters.

Note   It is possible to pass `input_shape=(...)` as a parameter in the first layer of a neural network to avoid the usage of the `Input` layer. While stylistic choices vary by person, it is generally better to use the `Input` layer rather than `input_shape=(...)` because it is clearer in terms of code organization. Moreover, the `Input` layer offers more options for handling data (see Keras documentation).

You may find these other layers generally helpful to add to neural network architectures, regardless of the specific application, to add more complexity or stability; the addition of these layers usually increases network performance (Table 1-2).

Table 1-2. *Table of useful regularization and normalization layers. Convolutional layers will be discussed later*

Layer	Implementation	Description
Dropout	`keras.layers.` `Dropout(rate=0.1)`	The dropout layer randomly drops some percent of the connections between the layer before dropout and the layer afterward. This percent can be specified via the `rate` parameter. Despite being simple, dropout is one of the most effective regularization methods to prevent large neural networks from overfitting.

(continued)

Table 1-2. (*continued*)

Layer	Implementation	Description
Batch normalization	`keras.layers.` `BatchNormalization()`	Batch normalization normalizes the outputs to the layer before it such that the mean is 0 and the standard deviation is 1. It was initially conceived to address a phenomena termed Internal Covariate Shift, in which the distribution of inputs to hidden layers in a deep neural network changes erratically as the parameters of the model are updated. However, Internal Covariate Shift has been shown not to affect model performance much. Nevertheless, batch normalization significantly smooths the loss landscape (the terrain that the optimizer must navigate to minimize the loss), which seems to be the reason it aids model performance so much. Many modern neural networks employ batch normalization because of the large boost it gives to most models in most problems.

Gaussian noise	`keras.layers.`
	`GaussianNoise`
	`(stddev=1)`
		The Gaussian noise layer adds zero-centered Gaussian noise to the outputs of the layer before it. Like dropout, the Gaussian noise layer adds an element of noise to prevent the neural network from overfitting. The Gaussian noise layer provides a different form of noise, though. Generally, placing noise layers near the input allows the neural network to become more robust to noisy data. We'll explore how this layer and layers like this can be used effectively in Chapter 2 in the context of self-supervised learning.

Now that the input and the body have been built, we can construct the output. The model should output two numbers corresponding to the targets in the dummy dataset, *Label 1* and *Label 2*. Moreover, these two numbers should be between 0 and 1, which means that we should apply the sigmoid function to its output. The sigmoid function allows for the output to be bounded between 0 and 1, representing the probability the input has the label *1*. One final Dense layer allows us to define two outputs and a sigmoid activation function: model.add(Dense(2, activation='sigmoid')).

Thus, our complete small neural network is as follows (Listing 1-4).

Listing 1-4. Building the complete Sequential neural network

```
import keras
from keras.models import Sequential
from keras.layers import Input, Dense
model=Sequential()
model.add(Input((3,)))
model.add(Dense(10, activation='relu'))
model.add(Dense(5, activation='relu'))
model.add(Dense(2, activation='sigmoid'))
```

It should be noted that there are many ways of importing that different people may find more or less effective given their coding style. For instance, in the preceding example, layers were imported as from keras.layers import a, b, ... When constructing large neural networks that require many different layers, however, importing individual layers is not necessary. In this case, one can use import keras.layers as L and reference each layer as L.layer_name. Using this method, the neural network could be rewritten without needing to import each layer by name at the beginning as follows (Listing 1-5).

Listing 1-5. Rewriting method of imports to avoid needing to import all layers explicitly

```
import keras
from keras.models import Sequential
import keras.layers as L
model=Sequential()
```

```
model.add(L.Input((3,)))
model.add(L.Dense(10, activation='relu'))
model.add(L.Dense(5, activation='relu'))
model.add(L.Dense(2, activation='sigmoid'))
```

Note that this book will use L.Layer format for most cases, except when it is necessary to present each layer individually to explain their purpose, in which the from keras.layers import a, b syntax is used (this occurring mostly in earlier chapters).

Now that we've defined the architecture of the model with the input-body-output framework, we need to provide information on other aspects of the model before it can be fitted on the data.

Step 2: Compile

Now that we've defined the architecture of the model with the input-body-output framework, we need to provide information on other aspects of the model.

The .compile() method of Keras models requires three key arguments:

- *Loss function*: The loss function is a mathematical function that takes in the true labels and the predicted labels and outputs the corresponding error. The loss function quantifies the "character" of the error – whether it punishes certain types of errors disproportionately, for example.

- *Optimizer*: The optimizer determines how the neural network changes its weights to minimize the loss function. Different optimizers are better suited toward different loss landscapes; some optimizers may be better, for instance, in loss landscapes that are smooth and flat. Others may succeed in more jagged ones.

- *Metric(s)*: This parameter does not affect how the neural network is trained, but it lets Keras now which metrics you want to use to monitor the neural network's performance. By default, Keras will record and provide information about the loss (and validation loss if you provide a validation dataset). However, it can also provide other metrics of error to provide a more holistic understanding of the neural network's performance.

For instance, a simple model could use mean squared error as the loss function, the Adam optimizer (Adam is generally used as a default optimizer), and also provide information on the mean absolute error and the accuracy: model.compile(loss='binary_crossentropy', optimizer='adam', metrics=['mae','accuracy']).

A good way to understand the compiling step is to understand each of these three parameters in relation to the loss landscape. The loss landscape is the relationship between each of the parameters of the model and the corresponding loss a model with those parameters will obtain.

For instance, consider the following loss landscape. The goal of the neural network is to find the location in the loss landscape with the smallest loss value.

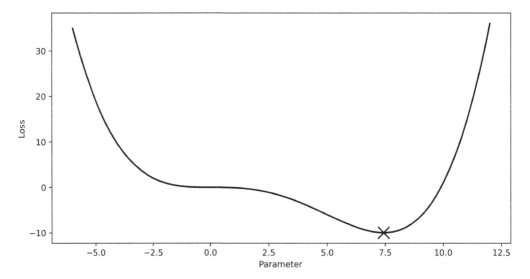

Figure 1-3. *Example loss landscape with the global minimum marked*

There are a few attributes that are common in most loss landscapes to note:

- There is a large range of feasible parameters. Parameters outside this range have a huge loss, which can be demonstrated by large sloping sides in which the loss increases dramatically. The feasible range, in this case, is approximately between –4.0 and 11.0. This means that a parameter value between this range results in a model that produces reasonable behavior with respect to the dataset.

- There is a global optimum/minimum marked with a cross. This is the parameter that minimizes the loss function and optimizes performance. In this example loss function, the parameter that optimizes the hypothetical model performance is about 7.5.

- There is a local minimum/plateau around the parameter range from –2.5 to 0.0. This is an area in which moving in either direction either increases error or leads to no increase in error. It can be difficult for a neural network to navigate these regions because the slope of the curve at that point does not give an indication of where the global minimum resides.

These features are common in loss landscapes, although in modern neural networks with millions or billions of parameters these landscapes are much more high-dimensional and difficult to interpret/visualize than in this example.

Let's get a better understanding of each of these elements of compiling a Keras model in relation to the concept of the loss landscape.

Loss Functions

The loss function is crucial to determining how the neural network arrives at its optimal solution. It defines the shape of the loss landscape by taking in the model predictions and the true labels and then returning a number to quantify the "goodness" of the model predictions.

The goal of designing a loss function is to best represent computationally or mathematically the features of a "best," "better," and "worse" model. A good loss function should be lowest where the model is the best. The loss should be lower if a set of model parameters is better and higher if a model is worse. Moreover, decreases and increases in the loss should be proportional to the "better-ness" or "worse-ness" of a model. If, for instance, we deem some behavior to be disproportionately more damaging to model performance than other behavior, our loss function should penalize that behavior disproportionately more to reflect our sense of what is better and worse.

For instance, the logic behind the mean squared error (MSE) loss, which calculates the average squared difference between the predicted label and the true label, is that large errors are disproportionately more damaging than small errors. The squaring mechanism forces any large errors to become especially prominent, and a model must address these large errors for the most significant reduction in loss before it focuses on smaller ones.

In many cases, optimizing the model on a different loss function results in a different solution (i.e., the set of parameters the final neural network has after training). There are several general loss functions that generally suffice for most problems and can be used to begin with. These are organized in Table 1-3.

It should be noted that some default loss functions can be mentioned explicitly by name by passing a string as an argument to the .compile() method, like .compile(loss='mse') or .compile(loss='mae'). However, if you want to use a more complex loss with specific parameters, pass a keras.losses object into the loss parameter rather than a string. For instance, if one wanted to use label smoothing (a method to prevent overconfidence) with binary cross-entropy loss, the model would be compiled as follows (Listing 1-6).

Listing 1-6. Example of passing a keras.losses object rather than a string as an argument when compiling. The parameter label_smoothing is abbreviated as l_smoothing for length

```
import keras.losses as L
model.compile(loss=L.BinaryCrossEntropy(l_smoothing=0.1),
              ...)
```

Even TensorFlow/Keras' default losses, however, are limited. TensorFlow Addons is a collection of community-contributed code that works with TensorFlow/Keras and builds upon its capabilities. TensorFlow Addons supports many more complex losses that are commonly used for certain types of problems. TensorFlow Addons can be installed via pip install tensorflow-addons and imported as import tensorflow_addons as tfa.

The Implementation Name column in Table 1-3 explicating helpful loss functions and use cases contains either a string ('loss_name'), a keras.losses object (keras. losses.loss_name()), or a tfa.losses object (tfa.losses.loss_name()). These can be passed into the loss parameter of the .compile command. Only a string is provided when there is no need for a keras.losses object because there are no parameters the user can provide.

Table 1-3. *A list of common loss functions for regression and classification problems*

Loss Function	Problem Type	Implementation Name	Description
Mean squared error	Regression	`'mse'`	The mean squared error is the averaged squared difference between the predicted label and the true label. MSE has the advantage of punishing larger errors disproportionately. For instance, if a model predicts an output of 1 and the true label was 3, the MSE would be 4. If the model increased its prediction by one unit to 2 instead of 1, the MSE would be 1. Increasing the model's prediction by one unit led to a fourfold decrease in error. As the model's prediction becomes closer to the true prediction, the MSE will decrease by less.
Mean absolute error	Regression	`'mae'`	The mean absolute error is perhaps the simplest loss function – it is simply the average distance the predicted value is from the actual value. It's almost never used in deep learning because of this simplicity, although it may suffice for certain simpler tasks and architectures.

(continued)

Table 1-3. (*continued*)

Loss Function	Problem Type	Implementation Name	Description

Huber loss	Regression	`'huber'`
		(default delta value is 1.0)
		`keras.losses.huber`
		`(y_true, y_pred,`
		`delta=1.0)`
			Huber loss is designed to be a combination of MSE and MAE, with the usage of a delta parameter. When the delta parameter equals 0, the Huber loss is equal to MAE; when it equals infinity, the Huber loss is equal to MSE. When the data is highly varied and has many outliers, use a low delta value; when the data is less varied, use a higher delta value.

Mean squared logarithmic error	Regression	`'msle'`	Mean squared logarithmic error is the mean squared difference between the logarithm of the true values and the logarithm of the predicted values. It is concerned with the ratio between the true and predicted values; it measures the percent-wise difference between target and prediction. MSLE especially penalizes underestimates more than overestimates.

(*continued*)

Table 1-3. (*continued*)

Loss Function	Problem Type	Implementation Name	Description
Cosine similarity	Regression	`'cosine_similarity'`	Cosine similarity computes the cosine of the angle between two vectors, being the true y values and the predicted y values. Cosine similarity loss in Keras is implemented such that a cosine similarity loss of 1 means two vectors are directly opposite each other (least similar), a loss of −1 indicates the two vectors point in the exact same direction (most similar), and a loss of 0 indicates the two vectors are orthogonal, or perpendicular, to one another. Cosine similarity loss is used when the direction of vector representations of the prediction matters more than its magnitude.

(*continued*)

Table 1-3. (*continued*)

Loss Function	Problem Type	Implementation Name	Description
Cross-entropy loss/negative log likelihood	Classification	`keras.losses.Binary Crossentropy (y_true, y_pred)` for binary classification `keras.losses. Categorical Crossentropy (y_true, y_pred)` for classification with more than two labels. Labels are one-hot encoded. `keras.losses.Sparse Categorical Crossentropy (y_true, y_pred)` for classification with more than two labels. Labels are integers.	Cross-entropy loss is one of the most popular and default losses for classification. It is the negative average of the logarithm of corrected probabilities. That is, one first finds the corrected probabilities, meaning simply p if the prediction is correct and $1 - p$ if the prediction is incorrect. The higher the corrected probabilities, the better the performance. Taking the negative logarithm of these probabilities allows for low corrected probabilities to have a disproportionately higher error than high corrected probabilities. The mean of these logarithms is the result of cross-entropy. Utilizing probabilities and logarithms is also related to information theory and entropy as a measure of a distribution's uncertainty.

(*continued*)

Table 1-3. (*continued*)

Loss Function	Problem Type	Implementation Name	Description

Focal loss	Classification	`tfa.losses.`
		`SigmoidFocal`
		`CrossEntropy()`
			Focal loss is most often used in classification with a highly imbalanced dataset. It is an altered form of cross-entropy that places a higher weight on less common examples and a lesser weight on more common ones. Focal loss was designed for the task of dense object detection, in which there is a very large imbalance between a specific class and other classes, but it has many other use cases.

There are many other losses not covered in this table that can be useful for certain subsets of problems. They can be found in Keras/TensorFlow/Tensorflow-Addons documentation.

However, even Keras and TensorFlow/Tensorflow Addons are limited with their selection of loss functions. You can create *custom loss functions* to use functions that have not been implemented or to change existing loss functions for your own purposes in ways that cannot be done through existing methods.

Creating a custom loss function takes the general following form (Listing 1-7). We'll use custom loss functions in later chapters to solve problems that require specialized loss functions beyond the currently existing ones.

Listing 1-7. The general form of a custom loss function

```
import keras.backend as K
def custom_loss(y_true, y_pred):
    result = do_something(y_true, y_pred)
    return result
```

Note that because the loss function is dealing with Keras tensors, it's easiest to use Keras backend or `tf.math` functions to perform mathematical operations, like `K.sum()` or `K.square()`. For instance, consider the following custom implementation of the mean squared error (Listing 1-8).

Listing 1-8. Example of writing the mean squared error loss as a custom loss function

```
import keras.backend as K
def custom_mse(y_true, y_pred):
    return K.mean(K.square(y_pred-y_true))
```

If the loss function is more complex, though, it can be difficult to write loss functions with only backend functions. In this case, TensorFlow's `.py_function()` allows for Python functions to be written as TensorFlow graphs that can be utilized as loss functions in neural networks. For instance, we can write `if/else` relationships in our loss function that otherwise would need to be written with TensorFlow functions like `tf.cond()`, which allows for condition statements but is difficult to work with and unintuitive (Listing 1-9).

Using TensorFlow's `.py_function()` to write Python operations into loss functions involves at least two functions:

- A Python function that handles the Python operations, like if/else/ elif, for loops, etc. It must accept a list of `tf.Tensor` inputs and output a list of tensors.

- A function that formally functions as the "real" loss function that is passed into the `loss` parameter of compiling. This function operates completely using TensorFlow/Keras objects and operations.

Listing 1-9. One example of the structure of writing Python functions as TensorFlow operations capable of functioning as a loss

```
def py_func(y_true, y_pred):
    if condition():
        return something(y_true, y_pred)
    else:
        return something_else(y_true, y_pred)
```

```
def custom_loss(y_true, y_pred):
    result = tf.py_function(func=py_func,
                            inp=[y_true, y_pred],
                            Tout=tf.float32)

    return result
```

The Tout parameter in py_function helps to specify the type of the output of the Python function. In general, defining custom losses is associated with type errors; these can generally be addressed by putting var = tf.cast(var, tf.float32) (or some other data type, like tf.float64) at the beginning of functions such that all input parameters to that function conform to a certain data type.

Additionally, your custom loss function needs to be able to handle batched data. One way of accommodating batches is to use the following structure: [func(batch) for batch in y_true]. This applies an operation to each batch.

It's also worth noting that loss functions cannot involve functions from external libraries unless they are converted into Keras/TensorFlow-adaptable functions, because loss functions need to be differentiable and non-Keras/TensorFlow functions are nondifferentiable. Your code might still run, but the model will not be able to access meaningful gradients and perform poorly.

In general, most custom loss function designs are simple enough to create that it's worth it to write up a custom loss function and reap the benefits of better performance. The loss function most directly determines the shape of your loss landscape, which controls which solution the neural network converges to. Choosing and/or adjusting the loss function to your particular problem is especially important for high performance.

Optimizers

While the loss function determines the shape of the loss landscape, the choice of optimizer determines how the neural network updates its weights to traverse the loss landscape to find its minimum. Which optimizer works best depends on the shape of the loss landscape, which is determined most directly by the loss function, as mentioned prior, but also by a variety of other factors, including the network architecture.

Like loss functions, when compiling, optimizers can be passed as string names, like 'adam', which initializes the default optimizer object or a TensorFlow/Keras optimizer object, like keras.optimizers.Adam(). TensorFlow Addons also has many optimizers not implemented in the core TensorFlow/Keras library that may be better suited for particular problems than the default available optimizers.

Almost all modern neural network optimizers are gradient-based, meaning that they rely upon the derivative, or slope, of the loss function at some point to judge which direction it should move in. This way, the optimizer can move in a "downward" direction, hopefully reaching the global minimum.

The primary problem with gradient-based methods is that loss landscapes of complex datasets and models have many local minima that the optimizer can get "stuck" in. Generally, optimizers are distinguished by how they handle local minima and plateaus. Some methods of dealing with these problems include

- *Momentum*: Imagine the optimizer as a ball rolling down a hilly landscape, trying to find the region of lowest elevation. Minima reside in the space between two hills. Once the ball reached a minimum, it would not immediately stop, because it has momentum from rolling down the hill – this momentum would allow it to continue rolling up the next hill for a little bit. If the next hill is too high, the ball will roll back down and oscillate until it reaches the minimum. If the next hill is not too high, the ball's momentum may allow it to roll over the next hill into potentially lower minima. Nesterov accelerated gradient (NAG) and other momentum-based optimization strategies use this logic to "jump out" of local minima.

- *Volatility of update*: Every step of fitting, the model evaluates the loss function and determines which direction to step in. If each update is highly volatile – it varies erratically and significantly – the optimizer can "jump" out of local minima simply because of this highly active behavior. However, this also means it risks jumping out of good local minima into worse solutions.

- *Adjusting learning rate*: While the derivative tells the optimizer which direction to step in, the learning rate decides how far the step should be. A high learning rate results in large step sizes, which can step over local minima that a low learning rate would have struggled to get out of. It should be noted, however, that if the model has found the true global minimum, setting a high learning rate runs the risk of jumping out of that global minimum.

See the following common optimizers and their implementation in a Keras model (Table 1-4).

Table 1-4. *A list of common less optimizers*

Optimizer	Keras Implementation	Description
Stochastic gradient descent (SGD)	`'sgd'` `keras.optimizers.SGD()`	Stochastic gradient descent performs an update to the weights for each training example, so it's faster than "vanilla" (standard) gradient descent. Because SGD updates weights frequently, training can be volatile. This means that SGD can jump and discover new local minima but also that it has difficult converging to a solution. SGD with Nesterov accelerated gradient performs well with shallow networks.
Adagrad	`'adagrad'` `keras.optimizers.Adagrad()`	Adagrad automates the adjustment of learning rates such that frequently occurring features are updated with a smaller learning rate (more caution) and less frequently occurring features are updated with a larger learning rate. Because of this automated learning rate adjustment, Adagrad is well suited toward sparse data. However, with time, the learning rate decreases steadily until it becomes functionally equal to 0, at which nothing new can be learned.

(continued)

Table 1-4. (*continued*)

Optimizer	Keras Implementation	Description
Root Mean Square Propagation (RMSProp)	`'rmsprop'` `keras.optimizers.RMSProp()`	RMSProp dampens Adagrad's problem of a continually decaying learning rate by decreasing the rate at which it decays. RMSProp is a common choice for deep neural networks because its method of decaying learning rate is effective in most loss landscapes.

Adaptive Moment Estimation (Adam)	`'adam'`
	`keras.optimizers.Adam()`
	See `tfa.optimizers.LazyAdam()` for sparse data.
	See `tfa.optimizers.RectifiedAdam()` () to address (rectify) the high variance of adaptive learning rates.
		Adam addresses Adagrad's aggressive learning rate problem in a similar way to RMSProp. However, Adam considers the gradient-update history, meaning that the update is not only based on the current gradient but also on the previous updates. We don't need to rely on the current gradient but can inform the update behavior based on several steps. Adam is commonly used in deep networks.

Choosing an appropriate loss function to optimize upon is essential toward good model performance.

Metrics

Metrics are not involved with changing or navigating the loss landscape. Rather, they help to understand the goodness of the model more holistically. Although the metric and the loss are structurally similar in that they take in the predicted label and the true label and output a number to quantify the error, they should be designed differently.

The purpose of a loss is to guide the optimizer to the best solution, whereas the purpose of a metric is to understand as a human if a model is good.

Thus, in designing a loss function, there is an emphasis on movement (i.e., guiding the optimizer to move toward an ideal solution), which can come at the cost of interpretability. That is, we care more about guiding the network than actually interpreting its performance during training. On the other hand, a good *metric* in the context of deep learning is one that, beyond the prerequisite of being suitable to your problem (problem and dataset type – classification/regression, balanced/imbalanced, etc.), must be easily interpretable. Ideally, metrics can provide a richer set of data either to support human decisions about what you should do with the model. For instance, if the model reaches a very low loss value *and* several other meaningfully different metrics confirm the model's high performance, you can trust that the model is not simply good in terms of the *loss function*, but generally across several metrics. A model that performs well both in their loss and in their metrics will likely succeed at modeling the *phenomena the dataset represents*, not only the dataset itself (i.e., overfitting).

If a model performs well in their loss but performs poorly with other metrics, there are two possibilities to consider:

- *Is the loss function representative of a "good solution"?* If the loss function is not well constructed, it may guide the neural network to a solution that minimizes the loss but does not perform well for other metrics.

- *Are the metrics relevant to the problem?* It may be that certain metrics are not relevant to the phenomenon the neural network is attempting to model. If you deem that this is the case, you can choose other metrics that are more relevant to the problem.

Comparing metric indicators to model performance can yield important insights into how the model is performing. Remember that deep learning and machine learning generally are concerned with modeling *phenomena*, or the source of data, not the actual dataset itself (although we must model the dataset to model the phenomena). That is, we want a model to separate images of cats and dogs, not to – as its end goal purpose – memorize an image dataset, even though in order to do so the model needs to learn representations in the image dataset. Metrics allow us to consider the model's performance and behavior more broadly and richly than only with the loss.

You can pass in many of the losses discussed prior as metrics, like mean squared error or cross-entropy. However, there are some additional metrics that cannot, should not, or are very difficult to be used as loss functions that are nevertheless valuable as metrics for monitoring network performance (displayed in Table 1-5).

Like loss functions and optimizers, in Keras metrics can be passed into the `metrics` parameter of compiling as a string, a `keras.metrics` object, or a `tfa.metrics` object.

Table 1-5. *A sample list of common metrics used in classification. More or less all loss functions for regression can be used as metrics*

Metric	Keras Name	Description
Accuracy	`keras.metrics.BinaryAccuracy()` for binary labels `keras.metrics.CategoricalAccuracy()` for categorical one-hot labels	Accuracy is one of the most basic metrics, calculated as the number of labels predicted correctly divided by the total number of labels.
Area Under the Curve (AUC)	`keras.metrics.AUC()`	The AUC (or AUROC, or Area Under Receiver Operating Curve) is a metric used for binary classification that ranges from 0 to 1. The AUC indicates the probability that a positive drawn uniform randomly has a model output higher than a negative drawn uniform randomly.
F1 score	`tfa.metrics.F1Score`	The F1 score is the harmonic mean of the precision and the recall, ranging from 0 to 1. Functionally, it is similar to accuracy in balancing consideration of the correctness and incorrectness of the model but works better with imbalanced datasets.

One can also create a custom metric using the same methods as creating a custom loss function. It's good to build custom metrics to suit your particular problem, since general metrics often have weaknesses that may be misleading on datasets with certain attributes, like being imbalanced or having many outliers. However, make sure that custom metrics are interpretable.

Step 3: Fit and Evaluate

Now that the model architecture has been defined and the model has been compiled, the model can be fitted to the data. Fitting generally takes the following form: model.fit(x=x_data, y=y_data, epochs=30, callbacks=[callback1, callback2]).

In this case, x_data and y_data are provided to fit the model. These can take the form of numpy arrays or pandas DataFrames. However, for many large datasets, for instance, of images, using these explicit methods often causes memory problems. Other more efficient data flow methods will be discussed later.

The epochs parameter refers to the number of times the model will loop over the entire provided dataset. If epochs=5, for instance, the model will run over the entire dataset five times.

Although callbacks are not required for fitting a model, it is good practice to use them. Callbacks are operations performed after each epoch and therefore allow for recording or changing of the model during its training. Three callbacks are commonly used – save model weights, early stopping, and slowing learning rate on plateau.

The ModelCheckpoint callback (Listing 1-10) saves model weights after each epoch to a specified filepath; if save_best_only is set to True, only the best set of model weights is stored to the location. Which weights are best can be specified by the argument for the monitor, which specifies the metric or loss to monitor, and mode, which specifies whether the best model is the one that minimizes or maximizes the metric or loss. For instance, if monitor is set to 'accuracy' (and accuracy is a metric provided during compiling), mode should be 'max'.

Listing 1-10. Model checkpoint callback syntax

```
import keras.callbacks as C
mc = C.ModelCheckpoint(filepath='output/folder',
                       save_weights_only=True,
```

```
                    monitor='val_loss',
                    mode='min',
                    save_best_only=True)
```

The ModelCheckpoint callback allows for the model to stop training once it stops improving. This can help conserve computing resources. The monitor parameter specifies the metric or loss that will be monitored for improvement. A change in the monitored metric or loss less than min_delta will be counted as no improvement. By default, min_delta will be counted as 0. Lastly, patience determines how many epochs of no improvement are needed for the model to stop training. It can be used in Keras as such: es = C.EarlyStopping(monitor="val_loss", min_delta=0, patience=3).

The ReduceLROnPlateau callback also checks if the model stops improving (it hits a "plateau") but lowers the learning rate instead of stopping training, as EarlyStopping does. The monitor and patience parameters play the same role as EarlyStopping does; the ReduceLROnPlateau callback has another parameter, factor, which determines the factor by which the learning rate is multiplied by every time when the model is determined to have stopped improving. It can be used in Keras as such: rp = C.ReduceLROnPlateau(monitor="val_loss", patience=3, factor=0.1).

Running the .fit() method for this neural network on the dummy dataset for 1000 epochs yields a progress output as displayed. The value of the loss function and metrics are displayed in the progress output (Listing 1-11).

Listing 1-11. Example output for a neural network on the sample dataset when the model is fitted. Note the decreasing loss just in the first three epochs. Additionally, because our dummy dataset is so small, it is counted only as one batch. Batches are groups of data that are processed simultaneously to better inform gradient updates. With larger datasets, the progress bar will indicate how many batches have been processed in the current epoch

```
Epoch 1/1000
1/1 [==============================] - 1s 610ms/step - loss: 0.6230 - mae:
0.4616 - accuracy: 1.0000
Epoch 2/1000
1/1 [==============================] - 0s 4ms/step - loss: 0.6209 - mae:
0.4604 - accuracy: 1.0000
Epoch 3/1000
```

```
1/1 [==============================] - 0s 2ms/step - loss: 0.6188 - mae:
0.4592 - accuracy: 1.0000
```

While these are good indicators of the model's performance, the progress can be stored for analysis and visualization by assigning the output of model.fit() to a variable, like such: history = model.fit(...).

history.history is a dictionary in which each key is the name of a loss or metric, like 'loss', 'mae', and 'accuracy', and the corresponding value is an array of those values for each epoch. The loss can be visualized, for instance, as follows (Listing 1-12, Figure 1-4).

Listing 1-12. Plotting the history

```
import matplotlib.pyplot as plt
plt.plot(history.history['loss'])
plt.xlabel('Epochs')
plt.ylabel('Loss')
plt.show()
```

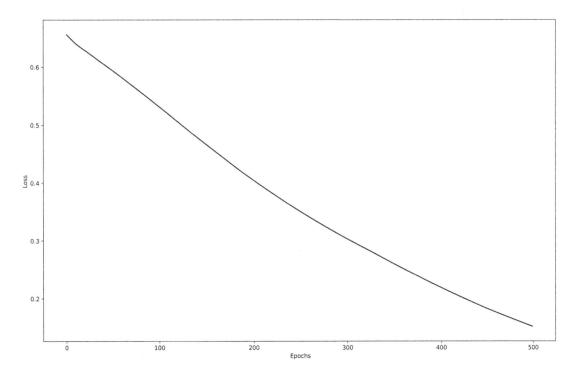

Figure 1-4. *Training performance of example model over epochs*

Visualizations can yield insights into training behavior, as well as future training decisions. The model seems to decline in loss rather steadily and consistently, although it is decreasing in speed. It is thus likely that a few more epochs of training can lead to a marginal decrease in loss. You can call `model.fit(...)` again and the model will continue fitting from where fitting terminated prior.

Model predictions can be computed with `model.predict(X_sample)`, and the model's performance on validation data can be computed with `model.evaluate(X_test, y_ test)`. Evaluations compute the loss and metrics passed in the compiling step for the provided dataset.

Based on the results of training, the model architecture and the parameters of training specified in the compiling step can be adjusted to optimize performance. Chapter 5 will discuss methods to automate this optimization.

Visualizing Model Architectures

Large and complex neural network architectures can be difficult to conceptualize when building. Keras provides a helpful utility to visualize the architecture of models that is invaluable when constructing neural networks. We can begin by plotting the architecture of the model we just built (Listing 1-13, Figure 1-5).

Listing 1-13. Plotting an example model with Keras

```
from keras.utils import plot_model
plot_model(model)
```

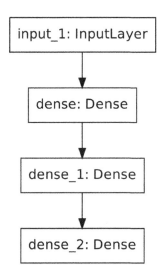

Figure 1-5. *Output of default* plot_model *command*

Keras visualizes each layer as a rectangle; each layer is annotated in the form "name: layer type." Keras automatically associates each layer with a name if the user needs to reference a specific layer. We can get more information about the model, however, by enabling Keras to show each layer's shape transformation and data type: plot_model(model, show_shapes=True, show_dtype=True). The output is visualized in Figure 1-6.

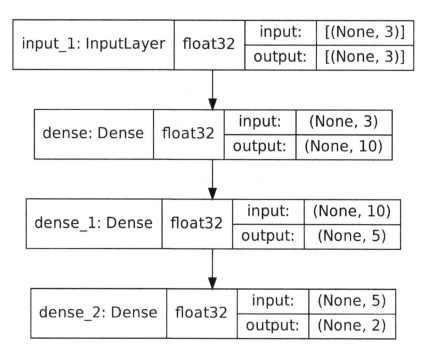

Figure 1-6. *Output of* plot_model *command with data type and shapes displayed*

The data type for all four layers in this example is float32. Knowing the data type allows you to debug any data type problems you may encounter with more complex data types and flows.

Knowing how each layer transforms the shape of the data is especially valuable, especially when working with high-dimensional data like images or text that require complex transformations. By looking at the diagram, for instance, we can see that the "dense_1" layer takes in data with a shape of (None, 10) and outputs data with a shape of (None, 5). Thus, this layer projects ten-dimensional data into a five-dimensional space.

> **Note**   The word "dimension" means many things in deep learning. In the traditional sense, a dataset with five features would be considered to be "five-dimensional." However, in image data, for instance, there are often three *spatial* dimensions, in that each image has a shape like (a, b, 3). Even though the data has three spatial dimensions, the neural network is technically operating in a*b*3-dimensional space. The term "dimensions" alone will refer to the number of elements in each data instance without regard for how they are organized, whereas the term "spatial dimensions" refers to the number of axes along which elements of each data instance are organized.

We will see examples in which visualizing models is helpful both immediately with nonlinear topologies and in Chapter 3 with autoencoder design.

Lastly, if you need high-resolution architecture diagrams, for example, in case of very large architectures that make viewing specific components more difficult, use the to_file='path/image.png' and dpi parameters. The latter controls the dots per inch; 200 to 400 is usually good for a clear and crisp image.

Functional API

The Sequential model adds layers sequentially via the .add() method. However, this type of model is limited in many respects, and more complex neural network structures use the Functional API.

In the Functional API, individual layers are defined as functions of the previous one (Listing 1-14).

Listing 1-14. General syntax of the Functional API

```
import keras.layers as L
input_layer = L.Input(shape)
second_layer = L.Layer_Name(params)(input_layer)
third_layer = L.Layer_Name(params)(second_layer)
output_layer = L.Layer_Name(params)(third_layer)
```

Each layer begins as its own variable, and each layer is initialized in relation to another layer (except for the first input layer), unlike the `Sequential` model, in which layers are defined in code in direct relationship to a `keras.models` object. It's important to note breakdown that this "double parentheses notation," as it is sometimes known to beginners, is two parts – the first set of parentheses is part of Python notation for initializing the layer object, and the second set is used to pass the previous layer into this initialized layer object. You don't have to perform both of these operations together (sometimes you cannot because the implementation requires separation, but generally it's easier to chain them together); for instance, you could write curr_layer = L.Layer_Name(params) to initialize some current layer on one line and write curr_layer = curr_layer(prev_layer) to connect it with the previous layer.

Once each of the layers are defined functionally, they need to be aggregated into a model object: model = keras.Model(inputs=input_layer, outputs=output_layer).

The benefit of assigning each layer to a unique and appropriately named variable is that they can then be referenced individually later. However, if this is not needed, it is common to redefine the layer each time with a variable like x. This builds the neural network functionally just as the prior method, but without the ability to reference individual layers (Listing 1-15).

Listing 1-15. Defining architectures functionally with repeated variable assignment for simplicity

```
import keras.layers as L
input_layer = L.Input(shape)
x = L.Layer_Name(params)(input_layer)
x = L.Layer_Name(params)(x)
x = L.Layer_Name(params)(x)
model = keras.Model(inputs=input_layer, outputs=x)
```

This variable redefinition notation can be confusing. Let's break it down: in line 3 of Listing 1-15, x is defined as a layer connected to the input layer. In line 4, x is defined as a layer connected to x; however, the object L.Layer_Name(params)(x) is initialized and connected before the variable x in line 4 is actually assigned. Thus, when the object is being initialized, x holds the previous layer defined in line 3. After the object is initialized and connected to the previously defined layer, the object itself is defined as x. In line 5, a new layer connected to this layer defined in line 4 is redefined as x, and so on. In line 6, x holds the last layer of the architecture and thus holds the output; all other intermediate layers besides the input layer have been lost in terms of being assigned to a variable (we can retrieve them by name; layer naming is discussed soon in this chapter, and non-variable layer retrieval methods are discussed in Chapter 2, where it is more relevant).

The process of training, predicting, and evaluating is the same for models based on the Functional API as the Sequential model once the model object is created. You can visualize this functionally defined model and compare it with the architecture of the sequentially defined model to verify their equivalence.

We will see more examples where the Functional API becomes extremely useful soon in the following chapters.

Translating a Sequential to a Functional Model

Earlier, we built the following model sequentially (Listing 1-16).

Listing 1-16. Previous Sequential model

```
from keras.models import Sequential
import keras.layers as L
model = Sequential()
model.add(L.Input((3,)))
model.add(L.Dense(10, activation='relu'))
model.add(L.Dense(5, activation='relu'))
model.add(L.Dense(2, activation='sigmoid'))
```

Now, we can build the model functionally without using the .add() method that Sequential models use (Listing 1-17).

Listing 1-17. Rewriting the Sequential model functionally

```
import keras.layers as L

# define all layers
input_layer = L.Input((3,))
dense_1 = L.Dense(10, activation='relu')(input_layer)
dense_2 = L.Dense(5, activation='relu')(dense_1)
```

```
output_layer = L.Dense(2, activation='sigmoid')(dense_2)

# aggregate into model
model = keras.Model(inputs=input_layer, outputs=output_layer)
```

We could also write the architecture of the sequentially defined model by using the variable redefinition method (Listing 1-18). Note that because the second step of aggregating the layers into the model requires an input and output layer, we cannot define the input layer as x.

Listing 1-18. Rewriting the Sequential model functionally with repeated variable assignment

```
import keras.layers as L

# define all layers
input_layer = L.Input((3,))
x = L.Dense(10, activation='relu')(input_layer)
x = L.Dense(5, activation='relu')(x)
x = L.Dense(2, activation='sigmoid')(x)

# aggregate into model
model = keras.Model(inputs=input_layer, outputs=x)
```

Building Nonlinear Topologies

In the prior section, we translated/rewrote a sequentially defined model into a functionally defined model. The Functional API is especially valuable, though, with nonlinear topologies, which are neural network architectures that are not and cannot be sequentially defined. Nonlinear topologies use *copying*, in which the output of one layer is copied and sent to two or more separate layers, and *concatenating* or other forms of merging, in which one layer's input is an aggregation of multiple layers before it. Contrast this with linear, sequentially defined topologies, in which each layer's input is only one layer's output and one layer's output is passed to only one layer's input.

Nonlinear topologies are valuable because they allow for more complex and differentiated representations of data. For instance, let's say that we want to train a neural network to classify if an image is of a cat or a dog. If we had trained a sequentially built convolutional neural network, the image would be passed through a set filter, and the extracted features from that one form of filter would be passed to the next layer. That is, each layer is limited by the one set of parameters determining how the prior character extracted information. You can find a more detailed and technical discussion of architectural nonlinearity in Chapter 6.

With a nonlinear topology in this example, we could pass the input through three separate filters to extract three different meaningful representations and extractions of the image:

- *Small size filter*: This layer captures smaller details, like the shape of the nose.

- *Medium size filter*: This layer captures larger details, like the textures of the fur or facial structure.

- *Large size filter*: This layer captures macro-scale patterns, like the shape of the animal body.

Although sequentially built models are said to go through these same steps (i.e., can develop their own "large," "medium," and "small" filter internal representations), they are more limited and constrained by their topology. Almost all modern deep neural network designs use nonlinear topologies of some sort.

Building nonlinear topologies with the Functional API is very intuitive. Let's work backward, beginning from a visualization of a topology (Figure 1-7) and writing code using the Functional API to create the model.

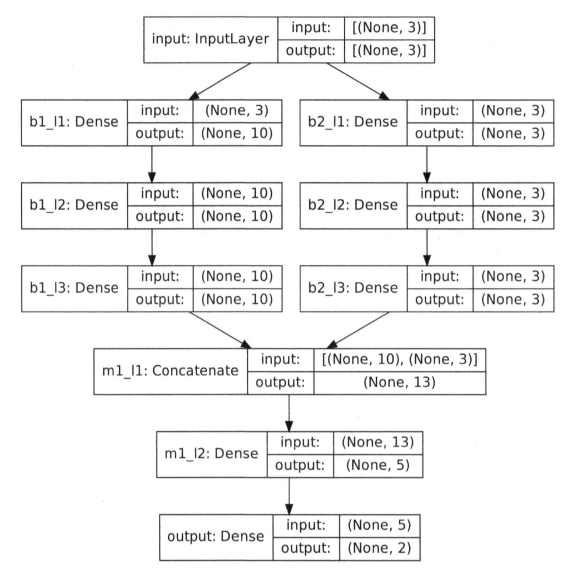

Figure 1-7. *Sample diagram of nonlinear topology to be reconstructed*

It should be noted that the diagram follows a naming procedure:

- b (branch): This is the series of layers after it is copied and before it is merged. Thus, "b1" indicates "branch1."

- l (layer): Thus, "l1" indicates "layer1."

- m (merging): This is the series of layers during and after the merging of branches.

This naming procedure, of course, is arbitrary and cannot handle more complex architectures, like if there are multiple merging sites. It's best to establish a naming procedure when working with nonlinear topologies for organization.

Let's begin by defining the input layer (Listing 1-19). Although Keras automatically names layers for you, it is good practice in complex topologies to implement your own naming procedure, which can be done by passing name='name' into each layer. You can also change the name after the layer is initialized by calling layer._name = 'name'.

Listing 1-19. Defining the input of a nonlinear functional model and specifying the name of layers

```
import keras.layers as L
input_layer = L.Input((3,), name='input')
```

The first branch can be constructed as such (Listing 1-20).

Listing 1-20. Defining one branch of a nonlinear functional model

```
b1_l1 = L.Dense(10, name='b1_l1')(input_layer)
b1_l2 = L.Dense(10, name='b1_l2')(b1_l1)
b1_l3 = L.Dense(10, name='b1_l3')(b1_l2)
```

The second branch can similarly be constructed (Listing 1-21).

Listing 1-21. Defining another branch of a nonlinear functional model

```
b2_l1 = L.Dense(3, name='b2_l1')(input_layer)
b2_l2 = L.Dense(3, name='b2_l2')(b2_l1)
b2_l3 = L.Dense(3, name='b2_l3')(b2_l2)
```

Now that the two branches have been created, we need to implement a merging. Merging layers take the form merging_layer()([l1,l2,...]), where l1, l2, and so on are prior layers whose outputs are being merged.

Keras offers many methods of merging:

- Concatenate(): The outputs of the previous layers are concatenated next to each other. For instance, if one layer outputs [0,1] and the other outputs [2,3,4], concatenating would yield [0,1,2,3,4].

- `Average()`: The outputs of the previous layers are averaged. For instance, if one layer outputs [0,1] and the other outputs [2,3], averaging would yield [1,2]. Unlike concatenating, in averaging all outputs must be of the same length.

- `Maximum()`: Returns the maximum for each corresponding element of the outputs of previous layers. For instance, if one layer outputs [0,5] and the other outputs [4,2], taking the maximum would yield [4,5]. Like averaging, in taking the maximum all outputs must be of the same length.

Keras also has similar `Minimum()`, `Add()`, `Subtract()`, `Multiply()`, and `Dot()` layers. Generally, the `Concatenate()` layer is used because it can handle previous layers of different lengths and because it preserves all their elements. The network can be trained to automatically perform its own aggregations after concatenation. However, in cases where you have a strong idea of what role you want certain branches or merging sites to perform, using other merging methods may be more successful than trusting the model to discover such relationships on its own with concatenation.

Note   If working with image data, make sure that all the feature maps being merged are the same dimension (e.g., you can't merge a convolutional layer with output shape (100, 100) and another with output shape (105, 105)). For concatenation-style merging of image data, Keras performs a depth-wise concatenation by default, in which the feature maps are stacked along the depth/ filters/channels axis. If feature maps with shape (100, 100, 32) are concatenated with default parameters to feature maps with shape (100, 100, 64), the concatenated output has shape (100, 100, 96).

The diagram demonstrates that the method of merging for this network is concatenation. We can finish the remainder of the network after the branches have been merged (Listing 1-22).

Listing 1-22. Merging and completing the nonlinear model

```
m1_l1 = L.Concatenate(name='m1_l1')([b1_l3, b2_l3])
m1_l2 = L.Dense(5, name='m1_l2')(m1_l1)
output_layer = L.Dense(2, name='output')(m1_l2)
```

The finished model can be aggregated into a `Model` object: model = keras. Model(inputs=input_layer, outputs=output_layer).

Throughout this book, we'll use a similar logic of first conceptualizing the network architectures and then using the Functional API to implement it. The simple design of the Functional API allows for building complex nonlinear topologies with for loops and conditionals. We will see a demonstration of this in later chapters.

Dealing with Data

Earlier, it was mentioned that training with `x_data` and `y_data` as `numpy` arrays or `pandas` DataFrames was possible, but usually too inefficient for most data types that would require deep learning applications. There are a few key signs or reasons why one would want to utilize alternative data flows to brute-force explicit data loading:

- Loading the data via an explicit brute-force method takes too much time.

- The raw data is in a complex enough form that writing a script to reform it into the desired formats would be arduous, and there are easier alternatives.

- The model runs into an OOM (out of memory) error.

- The raw dataset physically cannot fit into the memory allotted.

Because most deep learning applications will involve at least one of these problems, it's usually best practice to use alternative data flows to explicit data loading. There are many data flow methods, though, and it's important to choose one that fits your particular problem. There is no universal best solution for all problems, and choosing a specialized data flow method that is not suitable for your particular problem can be worse than just writing an explicit script.

> **Note**   The term "explicit" is used to refer to methods that very explicitly load
> the data as it is, without any special formatting or compression adapted to the
> data structure for efficiency. Sometimes, explicit methods are the best solution,
> especially with smaller datasets, in which setting up data flows may require more
> time investment than they return in value.

TensorFlow Dataset from Loaded Data

If you already have your dataset loaded into memory, there's a simple solution to put the data into the form of a TensorFlow dataset. This allows for the data to be processed more efficiently than in raw form. It can address OOM errors that you may encounter when using the raw data to train a model.

Note that we will need to import TensorFlow here to take advantage of its backend capabilities.

Say, you have two numpy arrays, X_train and y_train; you can use the from_tensor_slices function to convert these two numpy arrays into a TensorFlow dataset (Listing 1-23).

Listing 1-23. Constructing a TensorFlow dataset from already loaded arrays

```
import tensorflow as tf
from tf.data import Dataset as d
train_data = d.from_tensor_slices((X_train, y_train))
```

Note that the *x* and *y* data are grouped together into one dataset. When passing a TensorFlow dataset object into a Keras model for training, only one can be used, not two.

It's good practice to shuffle the dataset with train_data.shuffle(l), where l is the length of the dataset. This means that the order with which the network sees the data – which can influence its learning – is not influenced by how you loaded the data. If a cat/dog classification dataset is organized such that the first 50% of dataset items are all images of cats and the last 50% of items are all images of dogs, for instance, the network will learn poorly because it can't directly differentiate images of dogs and cats without both classes being present in close proximity to one another. Additionally, before training, batch the dataset with train_data.batch(n), where each batch has n data instances. This is not an optional step – failing to do so will yield an error in fitting.

When fitting the Keras model, the model can be correspondingly fitted by passing in the dataset: model.fit(train_data, epochs=10, …).

There are many methods of loading data directly into TensorFlow dataset from arrays, like text files or image files organized in a directory, which will be discussed next.

TensorFlow Dataset from Image Files

If you are working with an image dataset, it is likely arranged in a directory. It is not necessary to manually use another library like cv2 or matplotlib to load the images into memory and then to convert them into a TensorFlow dataset format with the prior method. TensorFlow allows us to automatically read images and convert them into elements of a TensorFlow dataset in one move with .map, which allows us to apply a function to each element of the dataset.

The idea is to have a dataset in which x is the filename of an image and y is the true label. Then, a parsing function will be applied to the dataset such that each filename is replaced with its corresponding image: unparsed_data = d.from_tensor_slices((filenames, labels)).

Let us define a function that takes in the filenames and outputs their image. It follows the following steps:

1. Read the content of the file.

2. Decode the file format (jpeg, png, etc.).

3. Convert to float values.

4. Resize the image to the desired size.

Luckily, TensorFlow has all of these operations implemented (Listing 1-24).

Listing 1-24. Creating a function to parse a TensorFlow dataset with filenames and labels

```
def parse_files(filename, label):
    raw_image = tf.io.read_file(filename)
    image = tf.image.decode_jpeg(raw_image, channels=3)
    image = tf.image.convert_image_dtype(image, tf.float32)
    image = tf.image.resize(image, [512,512])
    return image, label
```

43

A few notes about this function:

- You can use decode_png or another decoding function for particular image formats in place of decode_jpeg.

- Note that the label parameter is passed in and out of the function untouched. We still need to include this parameter because the dataset contains the label, which needs to be accounted for in a function that maps over the entire dataset.

- After resizing and before returning the image and label, you can also add several out of the many tf.image functions to perform image preprocessing, like adding random changes, flipping, cropping, adjusting saturation, brightness, and hue, and more.

This function can be mapped to the unparsed dataset and then batched, shuffled, and used for training: parsed_data = unparsed_data.map(parse_files).

You can use the logic of mapping a function upon a TensorFlow dataset to load and organize other data beyond images. Using a mapping function allows for a high level of control in organizing the data into some desired format.

TensorFlow datasets are commonly used for complex data of all types because of their efficiency and ease of usage.

Automatic Image Dataset from Directory

Keras provides a helpful function, .image_dataset_from_directory, which automatically constructs a tf.data.Dataset based on data organized in a directory where images are put into a folder of the corresponding class (see Listing 1-25). You can use Python libraries like os and cv2 to automate rearrangement of data into this format if it is not already.

Listing 1-25. Directory structure for Keras' automated creation of a TensorFlow dataset from a directory

```
data/
... class_1/
...... img_01.png
...... img_02.png
```

```
... class_2/
...... img_50.png
...... img_51.png
```

The following code can be used to quickly convert the directory into an image dataset (Listing 1-26). The `directory` parameter passes a string with the name of the directory directly containing the subfolders corresponding to each class.

Listing 1-26. Creating an image TensorFlow dataset from a directory

```
from keras.preprocessing import image_dataset_from_directory
train_data = image_dataset_from_directory(
    directory='data',
    batch_size=32,
    image_size=(256, 256)
)
```

The dataset can be then used like any standard TensorFlow dataset. However, since the data and labels are already zipped together, to apply transformations, the dataset needs to be unzipped.

ImageDataGenerator

A popular alternative to the TensorFlow dataset format is the `ImageDataGenerator` object, which generates image data in tensor form and augments them in real time rather than loading all data and generating image data first. Thus, `ImageDataGenerator` (Listing 1-27) is popular both because it deals with large datasets efficiently and because it continually offers new sources of random augmentation. Although TensorFlow datasets can also handle large data efficiently, it cannot add randomness to the data during training.

To use the generator, begin by initializing an `ImageDataGenerator` object. When initializing the object, provide it with information on what the augmentation or preprocessing for the images looks like. For instance, `rotation_range` takes in the degree range from which a random degree is selected and used to rotate the image. `width_shift_range` and `height_shift_range` randomly shift the image up or down a certain quantity of pixels. `horizontal_flip` randomly flips the image horizontally.

There are numerous other parameters that can be used to augment the image via brightening, normalization, shearing, zooming, and more – you can find specifics in the Keras documentation. If these available transformations aren't enough or don't suit your purposes, you can pass in a custom function to the preprocessing_function.

Listing 1-27. Setting up an Image Data Generator

```
from keras.preprocessing.image import ImageDataGenerator
data_gen = ImageDataGenerator(rotation_range = 20,
                              width_shift_range = 30,
                              height_shift_range = 30,
                              horizontal_flip = True,
                              preprocessing_function = func)
```

The generator acts like a mold through which data is transformed and shaped into a desired form. Note that the Image Data Generator doesn't actually generate new data by increasing the size of the dataset – it acts like an automated transformation that is run on the data every time before training. It technically is constantly generating new data, but not in a way that takes up more space; this makes it a popular form of augmentation in large dataset problems.

Now that you've specified the framework for how data should be changed, there are three methods through which data can be provided to the generator.

If your data is already loaded into array format but you want to take advantage of the ImageDataGenerator's real-time random augmentation capabilities, you can use the .flow() method, which takes in an *x* dataset and a *y* dataset and passes it through the specified parameters of the data generator. When fitting the model, the data generator functions as a dataset: model.fit(data_gen.flow(x_train, y_train), ...).

If the directory paths of images and their corresponding labels are in a DataFrame, you can use the .flow_from_dataframe method to "fill" the data with a source of data before it is passed as an argument into fitting (Listing 1-28).

Listing 1-28. Using the image data generator with data from a DataFrame to fit the model

```
data_gen_filled = data_gen.flow_from_dataframe(
    dataframe,
```

```
    x_col = 'filenames',
    y_col = 'label'
)
model.fit(data_gen_filled, ...)
```

If the images are organized into a directory where each subfolder contains all images of one class (like the format discussed earlier with TensorFlow datasets), you can use the .flow_from_directory method (Listing 1-29).

Listing 1-29. Using the image data generator with data from a directory to fit the model

```
data_gen_filled = data_gen.flow_from_directory(
    directory='data'
)
model.fit(data_gen_filled, ...)
```

Key Points

In this chapter, we discussed concepts and code to use Keras for implementing neural networks:

- Keras is simple to get started with and is extremely versatile.

- The Keras workflow has three steps: define model architecture, compile, and fit.

    - In the Sequential model, the model architecture is defined by stacking neural network layers upon each other. The input and output must be designed to accommodate the shape of your data.

    - The compiling step should take in at least three parameters: loss function, optimizer, and metrics. Each of these parameters deals with understanding the model in relation to the loss landscape. For each of these parameters, you can pass in a string, a Keras loss/optimizer/metrics object, or a TensorFlow Addons loss/optimizer/metric object.

- When fitting the model, provide the data, the callbacks, and the number of epochs to train for. You can collect and visualize the model performance to determine future steps.

- Keras offers utilities to visualize model architectures, which are valuable when constructing complex relationships between layers and/or analyzing how the shape of the data changes as it passes throughout the neural network.

- The Functional API allows you to define nonlinear topologies, which encourage the neural network to develop more complex representations of the data.

- Although you can pass data into the neural network as numpy arrays or pandas DataFrames, there are two more efficient formats that you can use: TensorFlow datasets and Keras Image Data Generators (for image data).

In the next chapter, we will build upon our knowledge of how Keras works to discuss how to take advantage of pretraining as a way to build complex, powerful, and successful models more easily and quickly.

CHAPTER 2

Pretraining Strategies and Transfer Learning

Steal from anywhere that resonates with inspiration or fuels your imagination... And don't bother concealing your thievery – celebrate if you feel like it. In any case, remember what Jean-Luc Godard *said: "It's not where you take things from – it's where you take them to."*

—Jim Jarmusch, American film director[1]

Obtaining a good deep neural network for most modern datasets is a difficult task. Optimizers need to traverse extraordinarily high-dimensional, jagged loss landscapes and differentiate between good and mediocre solutions using a limited set of tools. Hence, you will seldom see modern deep learning designs solving relatively well-studied problem types directly training the neural network on the data right after initialization – it's simply too difficult to get good results feasibly by training directly from scratch.

Part of the analytical creativity embedded within deep learning design, then, is a need to turn this difficult task – modeling a complex phenomenon with a deep neural network – into a more approachable and efficient process.

In this chapter, we'll discuss pretraining strategies and transfer learning: in essence, creatively stealing from knowledge contained in the dataset and in the weights of other models for to solve problems.

---

[1] *MovieMaker* Magazine #53 – Winter, January 22, 2004.

A. Ye, *Modern Deep Learning Design and Application Development,*
https://doi.org/10.1007/978-1-4842-7413-2_2

Developing Creative Training Structures

A model's *training structure* consists of the general "flow" with which it is trained – the dataset(s) it is trained on, the order of the entities it is trained on, how those datasets are derived, etc. Here, we are assuming the standard supervised mode of learning, in which the model is presented with a dataset consisting of input data (the x – what the model takes in) and the labels (the y – corresponding to what the model should output) (Figure 2-1). The model's *ultimate task* is to take in input data and to output the correct label; this dataset is the *task dataset*. Although we may train the model on other datasets and to perform other tasks, those auxiliary tasks are performed with the intent of helping the model succeed at the ultimate task. This distinction in terminology will prove to be helpful later.

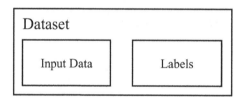

Figure 2-1. *Task dataset structure*

The simplest training structure is the initialize-train-complete flow that was discussed in Chapter 1 (Figure 2-2). The model begins training right after initialization, being fed the input data and predicting the labels. After training, we select a final model, which can just be the model after it has completed training or something like restoring the weights of the best performing model (something like Keras' `ModelCheckpoint` callback).

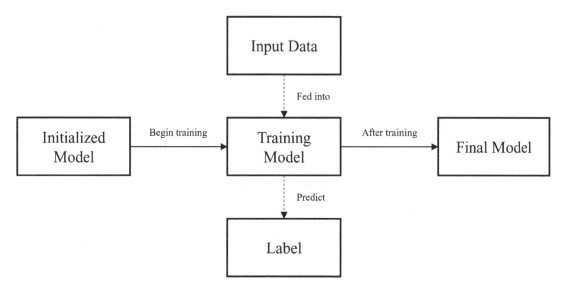

Figure 2-2. *Simple training structure of initialization and training*

Let's see how we can alter this simple training structure to improve training performance more easily.

The Power of Pretraining

The "pre" in "pretraining" indicates that it refers to a process that occurs *before* the formal training of the model on the dataset (Figure 2-3).

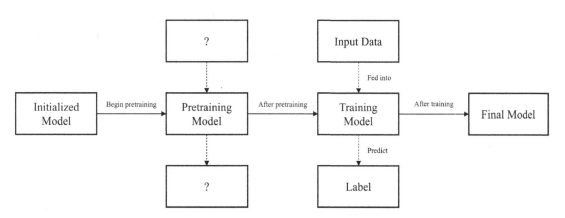

Figure 2-3. *Training structure representation of pretraining*

Given that we want pretraining to benefit training, pretraining should orient the model in a way that makes training "easier." With the pretraining step, the model's performance should be better than without pretraining. If pretraining makes model performance worse, you should reevaluate your methods of pretraining.

In pretraining, a model must be trained to perform a task different from its ultimate task of predicting the label based on the input data. This pretraining task ideally presents important "context" and skills for the ultimate task such that the model can attain better performance on that ultimate task.

Before we discuss the specific pretraining methods, we need to gain an intuitive understanding of what the vague terms "easier" and "better performance" mean; this will allow us to better understand how to use and design pretraining strategies. Pretraining is often discussed in relation to two of its primary advantages:

- *Time*: Pretraining can decrease the time needed for the neural network to converge to some solution.

- *Better metric score*: The model attains a metric score higher/lower (depending on the metric) than it would have without pretraining. For instance, it attains a lower cross-entropy score or a higher accuracy.

While these are two dominant attributes of pretraining, they are the result of an underlying phenomenon: conceptually, the process of pretraining brings the optimizer "closer" to the true solution in the loss landscape.

Consider the following loss landscape: it is the objective of the neural network optimizer to find the set of parameters such that the corresponding loss is minimized. This landscape has several features common in most modern loss landscapes: it has several local minima but only one global minimum and sloping, jagged fluctuations (Figure 2-4).

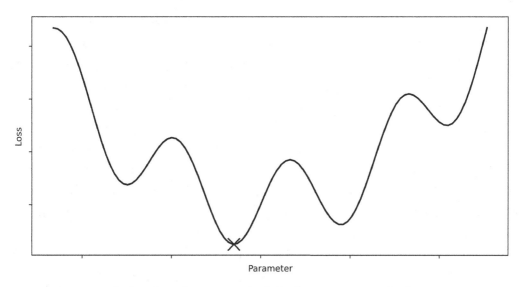

Figure 2-4. *Sample loss landscape with global minima marked*

A model without pretraining may initialize and follow the following path. Since the first minima it encounters is more or less shallow, say the optimization algorithm overcomes it and discovers the next minima. Since this minima resides in a deeper pit and the optimizer has travelled a long distance, it will likely judge this to be the global optimum and converge at that location (Figure 2-5).

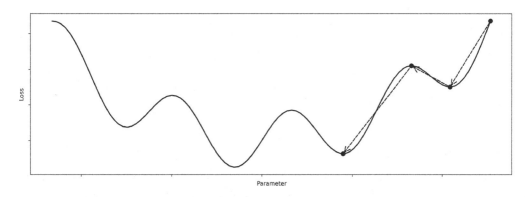

Figure 2-5. *Example optimizer movement without pretraining*

With pretraining, however, we're able to get the model "closer" to the true global minimum of the loss landscape such that it converges to a more optimal solution faster, because it already begins from a place "close" to the global optimum. Correspondingly, we can use a less risky or erratically behaving optimizer because it begins from a convenient position (Figure 2-6).

Keep in mind that pretraining does not necessarily bring the model "closer" to the optimum in the literal sense of distance but rather that it becomes more convenient or "easier" for the model to find the optimum than if it hadn't undergone pretraining. An optimizer very close to the global optimum but with several very deep local minima and very high local maxima would find it more difficult to arrive at the global optimum than an optimizer that was farther but whose path was smoother and easier to descend. What "closer" exactly means is dependent on the shape of the loss landscape, the optimizer's behavior, and a host of other factors. For simplicity, though, thinking of "closer" in terms of "how easy it would be to navigate to the global optimum" suffices.

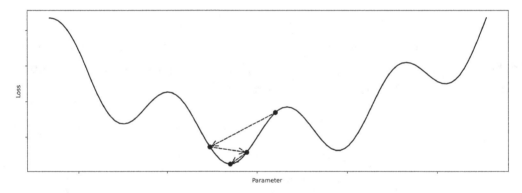

Figure 2-6. *Example optimizer movement with pretraining*

Note that the actual loss landscape for the pretraining task is different from that of the loss landscape of the ultimate task (as is displayed in Figure 2-6). It is not that pretraining operates within the loss landscape of the ultimate task and moves the model to a convenient position – that is just training on the task dataset, not pretraining.

Rather, in pretraining, we rely upon a similarity between the loss landscapes of the pretraining task and the ultimate task the model aims to perform. Therefore, a model that succeeds at the pretraining task should be at a generally successful location in the loss landscape for the ultimate task, from which it can further improve. This also means that pretraining may not be helpful if you believe the loss landscapes of the pretraining dataset and the task dataset are very "different." Because the two often cannot be compared quantitatively for technical reasons, it is up to you to decide the necessity and performance boost of pretraining in a particular context.

This visual depiction is intended to be more a conceptual method of understanding what pretraining intends to do to aid your designing and usage of pretraining strategies. It is, of course, not entirely representative of the high-dimensional spaces that modern neural network optimizers operate in or the technicalities of transfer learning loss landscape shapes and manipulations. For instance, it's unlikely that pretraining will get the model to a point where the optimizer encounters no local minima, as presented in Figure 2-6. Nevertheless, we can use this conceptual framework to justify and explicate the two key observed benefits of pretraining – speed and better metric scores.

- *Speed*: Pretraining brings the model "closer" to a solution that the optimizer would be "satisfied with" (could converge to), because it's already done much of the work. Moreover, most modern optimization strategies involve a decreasing learning rate. A neural network that has not had the benefit of pretraining needs to travel more (perhaps in distance, perhaps in overcoming obstacles like local minima or maxima) to near the true solution. By then, its learning rate can be expected to have decayed significantly, and it may not be able to get out of local minima. On the other hand, if a pretrained model begins already near the true solution, its learning rate begins "fresh" and undecayed; it can quickly overcome local minima that lie between it and the true solution.

- *Better metric scores*: Pretraining reduces the quantity of obstacles between a model and the true solution. Thus, the model is more likely to converge to the true solution. Moreover, as discussed prior, the optimizer's learning rate is more "fresh" near the true solution than an optimizer without pretraining and thus is less susceptible to being vulnerable to minor obstacles that may have tricked the optimizer without pretraining.

Next, we'll build upon this conceptual model to discuss the intuition behind two pretraining methods: transfer learning and self-supervised learning.

Transfer Learning Intuition

Transfer learning is premised upon the idea that knowledge gained from solving one problem can be used, or transferred, in solving another problem. Usually, the knowledge derived from a more general problem is used to aid a model's ability to address a more specific problem.

Consider two datasets: a general dataset and a task dataset. The general dataset is designated for pretraining, whereas it is the model's ultimate purpose to perform well on the task dataset (Figure 2-7).

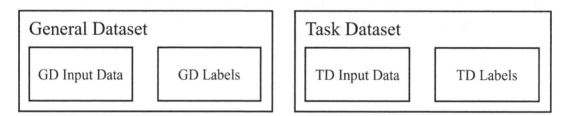

Figure 2-7. *General dataset and task dataset visual representations. GD = general dataset, TD = task dataset*

In transfer learning, an initialized model is first trained on the general dataset to provide context and skills required to succeed on the task dataset. After the model is pretrained on the general dataset, it is then trained on the task dataset. This training structure is visualized in Figure 2-8. Because the weights are retained from the end of pretraining to the beginning of pretraining, the model has already acquired important skills and representations for pretraining that it can use when training on the general dataset.

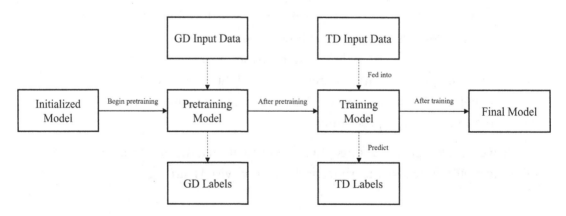

Figure 2-8. *Training structure visual representation of transfer learning*

Usually, the pretraining component of transfer learning is not done in house (by yourself). Selecting an appropriate general dataset and conceptualizing, implementing, and training a model to perform well on that general dataset is a significant amount of work, and it's not necessary. There is a repository of pretrained models available in most deep learning frameworks (Figure 2-9), like Keras/TensorFlow and PyTorch and other sources, like pypi libraries, online code forums, and hosting sites. Each of these pretrained models is trained on some general dataset and is available for you to use.

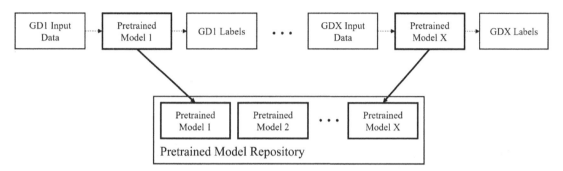

Figure 2-9. *Pretrained models in a model repository. GD = general dataset. GD1, GD2, ..., GDX indicate several different general datasets upon which several pretrained models are correspondingly trained. Note that in practice the relationship between general datasets and pretrained models is one to many (many pretrained models with different architectures are trained on the same general dataset)*

You can then choose one of the already pretrained models and begin training them on your specific task dataset (Figure 2-10).

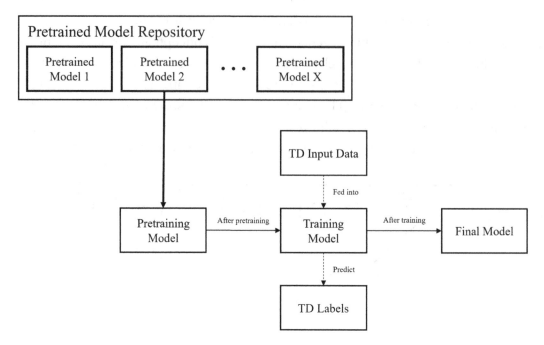

Figure 2-10. *Using models from a pretrained model repository to directly begin training on the task dataset*

The number of problems with which this set of pretrained models could be used vastly outnumbers the number of pretrained models in the repository. This is fine, though, because we can expect each pretrained model to be capable of being applied to a wide array of problem types, since each model is expected to possess a form of "general knowledge."

For instance, let's say that we want to train a model to classify images of dogs and cats (Figure 2-11). We decide we want to use transfer learning and choose a general dataset of all sorts of real-life items and objects, like a parrot, an airplane, a car, a tomato, a fish, etc. For the purposes of this example, let's say that images of dogs and cats are not present in the general dataset.

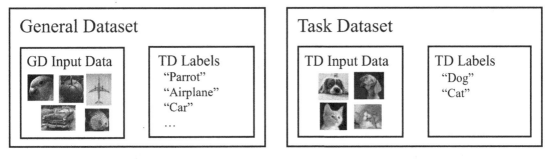

Figure 2-11. *Example task dataset and general dataset*

There may seem to be little connection between the general dataset and the task dataset. However, if we distance ourselves from instinctive human topic-based thinking – "this is a car, that is a dog; they belong to different categories (i.e., transportation and animals)" – we can see many similarities.

- *There are many edges*: Edges define the shapes of the airplane, car, and fish; likewise, they are important in defining the shape of a dog or cat's head.

- *Light is important*: Since both datasets are two-dimensional representations of three-dimensional objects, light is important because it helps determine the shape and contour of the object.

- *Texture matters*: Real-life objects often have similar shapes but are differentiated by the texture of their surface. Likewise, this seems to be important toward differentiating between images of dogs and cats.

A model that succeeds on the general dataset must have already adapted to and accounted for these features of the general dataset. In a sense, it "understands" how to "interpret" edges, the dynamics of light, texture, and other important qualities. All that is necessary afterward is to adapt those learned skills and representations toward a similar but more specific task.

You may ask, "why not approach task dataset directly? Why develop these auxiliary skills when the model could have developed those skills more specifically and directly for the task dataset?" Indeed, in some cases transfer learning is not a suitable choice. However, in most sufficiently difficult tasks, pretraining via transfer learning helps to set up the foundation for learning that would be difficult for the model to build by itself.

Imagine you are teaching a young child how to add two-digit numbers only through repeated examples. You have two general possible teaching strategies:

- Teach the child directly to add two-digit numbers by repeatedly showing them examples (e.g., 23+49=72).

- First, teach the child how to add one-digit numbers by repeatedly showing them examples (e.g., 3+9=12, 2+6=8). Then, teach the child to add two-digit numbers by repeatedly showing them examples (e.g., 23+49=72).

The latter is more likely to be a successful strategy, because the average length of the jump from knowing nothing about addition to adding one-digit numbers to adding two-digit numbers is much smaller than the average length of the jump from knowing nothing about addition to adding two-digit numbers. Transfer learning should operate by this same intuition of teaching the model a general task such that a more specific, ultimate task becomes easier to approach.

Self-Supervised Learning Intuition

Self-supervised learning follows the logic of pretraining – some sort of pretraining task is performed before training the model on the task dataset to orient it toward better attaining representations and skills needed to perform well on the ultimate task.

The difference between self-supervised learning and transfer learning is that in transfer learning, the pretraining dataset is different from the task dataset, whereas in self-supervised learning, the pretraining dataset is constructed from the input data of the task dataset. Thus, while you would need two datasets to build the complete transfer learning training structure, only one is technically needed to build the complete self-supervised learning training structure (Figure 2-12).

Note   Here, we are using a unique definition of what a "different" dataset constitutes. If dataset A can be derived completely from dataset B (for instance, by flipping the images or changing image color), for the purposes of this concept, the two datasets are not different, even though on a technical level there are training instances in one dataset that cannot be found in the other. On the other hand, if dataset A cannot be derived completely from dataset B (for instance, deriving ImageNet from CelebA), the two datasets are different. The main focus of what constitutes "difference" here is the informational content of the dataset, not the individual specific training instances. This allows us to distinguish transfer learning and self-supervised learning.

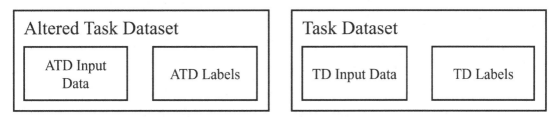

Figure 2-12. *Self-supervised learning datasets. ATD = altered task dataset*

It's important to note that the altered task dataset is generally derived only from the input data of the task dataset. For instance, if a model's task is to identify whether an image is of a dog or a cat, the altered task dataset could only be built upon the images (the input data of the task dataset), not the labels (the labels of the task dataset).

Designing altered task datasets is a fascinating demonstration of the creativity embedded within successful deep learning solutions. There are many examples of ways you can construct labels from the task dataset to form an altered task dataset:

- Add noise to some of your data and none to others. Train the model to classify to which data instances noise was added (this is a binary classification problem: noise/no noise). This can help the model better separate noise and data, allowing it to develop an underlying representation of what key features of the data should look like.

- Add varying degrees of noise to all instances in the dataset. Train the model classify the degree of noise (this is a regression problem). For instance, if you varied the standard deviation of Gaussian noise, the model would predict the standard deviation. This not only helps the model detect if there is noise, but to what extent it exists. While this is a more difficult pretraining task, the model would be encouraged to develop sophisticated representations of key features and structures of the data.

- Assume there is a dataset of colored images and color is important to the model's ultimate task. Convert the images to grayscale and train the model to construct a colorized image from the corresponding grayscale one (this is an image-to-image task). In this case, the altered task dataset input data is the grayscale image and the labels are the colorized images. With this self-supervised pretraining exercise, the model gains an understanding of what color certain objects should be.

- Assume there is a model that needs to perform the NLP (Natural Language Processing) task of text classification. Take the text samples (input data for the task dataset) and randomly hide one of the words. Train the model to predict what the hidden word is. For instance, the model would predict "cats" when given "it's raining ____ and dogs." This allows the model to gain an understanding of how words function in relation to each other bidirectionally – it needs to take advantage of information both before and after the hidden word. This method of self-supervised learning is commonly used in modern NLP architectures.

In each of these examples, none of the task dataset's labels are needed – only the input data, or the x, of the task dataset is used in constructing these supervised datasets. You, as the deep learning engineer, can make some change to the task dataset input data and construct labels from that change.

Once the altered task dataset has been derived from the input data of the task dataset, it can be used in the pretraining training structure (Figure 2-13).

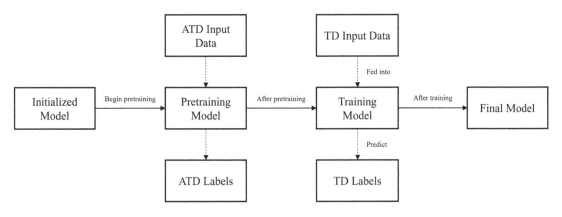

Figure 2-13. *Self-supervised learning training structure*

Note that in self-supervised learning (and pretraining in general), attaining a high performance metric is not the ultimate goal. In fact, if a model performs too well on the pretraining task, it may be that the pretraining task was too easy and hence did not support the growth of valuable skills and representations of data for formal training. On the other hand, if the model performs very poorly, the pretraining task may be too difficult.

It should be noted that self-supervised learning is technically unsupervised learning because we are extracting insights from the input data without any knowledge of the labels. However, self-supervised learning is a more commonly used and appropriate term: there is undoubtedly a supervised character to this sort of operation that distinguishes it from traditional unsupervised machine learning algorithms like K-means in that while the data generation (constructing altered task dataset) is technically unsupervised, the learning procedure (gradient updates, etc.) is supervised.

Self-supervised learning is also valuable because it allows us to capture valuable information without needing labels. Datamation estimates that the amount of unstructured data is increasing by 55% to 65% every year,[2] and the International Data Corporation projects that 80% of data will be unstructured by 2025.[3] Unstructured data is defined as data that cannot be easily fit into a standard database model. An auxiliary

---

[2] www.datamation.com/big-data/structured-vs-unstructured-data/.

[3] https://solutionsreview.com/data-management/80-percent-of-your-data-will-be-unstructured-in-five-years/.

characteristic of unstructured data, thus, is that there are seldom corresponding labels to train massive deep learning models with. However, the logic of self-supervised learning allows us to find hidden structures in unstructured data without labels and to exploit that knowledge to solve supervised problems.

For this reason, self-supervised learning is especially valuable for training on small datasets, in which there are few labels. Pretraining via self-supervised learning can allow the model to derive important insights it may not have obtained with only the traditional training structure.

Unlike transfer learning, there is no repository of self-supervised pretrained models, because in self-supervised learning, the pretraining task must be designed based on the task dataset. However, altering task datasets for self-supervised learning is generally not difficult if you have a good grasp of programming and data flows (see Chapter 1).

Yann LeCun provided an analog for self-supervised learning in how humans and animals learn at a 2020 AAAI (Association for the Advancement of Artificial Intelligence) conference:

Self-supervised learning ... is basically learning to fill in the blanks. Basically, it's the idea of learning to represent the world before learning a task. This is what babies and animals do. We run about the world, we learn how it works before we learn any task. Once we have good representations of the world, learning a task requires few trials and few samples.

—Yann LeCun, Chief AI Scientist at Facebook, speaking at AAAI 20[4]

Transfer learning and self-supervised learning both fall under the general strategy of pretraining and are similar in that the model engages in some supervised task before it is trained on the task dataset. However, the two pretraining methods are different not only in terms of technicalities (i.e., "different" datasets vs. "same" datasets) as discussed prior but also as a matter of outcome and what knowledge is developed.

- Transfer learning tends to develop "prediction *skills.*" That is, the weights from transfer learning are derived from the general dataset, which usually comes from a significantly different context than the task dataset (topic/content-wise). Much of the value of transfer learning is in the predictive skills it develops – learning how to recognize edges, to look for texture, to process color, etc. rather than actual knowledge of the content or topic of the dataset.

---

[4] www.youtube.com/watch?v=UX8OubxsY8w.

- Self-supervised learning tends to develop "world-representing knowledge." It builds the fundamental representations of the "world," or context, that the model will need to understand to perform well on the task dataset. The actual skill of predicting whether – for instance – noise was or was not added to a training instance may not be of much use, but the process of deriving that skill requires gaining an understanding of the data "world." Self-supervised learning allows for the model to get a glimpse into building fundamental representations and "feeling around" for the dataset's content and topics.

Of course, this is not to suggest that transfer learning does not develop world-representing knowledge or that self-supervised learning does not develop important skills. Rather, these are the root-level "spirits" or "characters" of transfer learning and self-supervised learning that can be used to guide intuition and what to expect from either method.

You may notice that we have been exploring these concepts through relatively fluid and artistic descriptors and ideas – "extracting insights," "developing representations," "world-representing." While a textbook may attempt to formulate these ideas in equations and mathematical relationships, the truth is that even the most modern deep learning knowledge cannot fully explain and understand the depth of neural network behavior. Having an intuitive grasp of how neural networks function and "learn" – even if it is not mathematically rigorous – is valuable toward successful and efficient design.

Next, we will explore practical concepts in pretraining to understand how to manipulate neural network architectures for the implementation of transfer learning and self-supervised learning.

Transfer Learning Practical Theory

Previously, we discussed conceptual frameworks to gain an intuitive understanding for what transfer learning is, what it does, and how it operates. However, because there are many "hidden" considerations when implementing transfer learning, in this section we will explore the more practical aspect of theory – concepts and ideas to aid your implementation of transfer learning.

The purpose of this section is not to provide specific code or discuss examples (that will be left to the next section), but instead to provide an introduction to important concepts for implementing pretraining, like the structure of pretrained models and how pretrained models are organized in Keras.

Transfer Learning Models and Model Structure

There are several important pretrained models and datasets to know. In this section, we will primarily be exploring image-based models, although you can apply much of the logic to other contexts. Image-based pretrained models generally follow a standard structure, consisting of two key steps: feature extraction and feature interpretation (Figure 2-14).

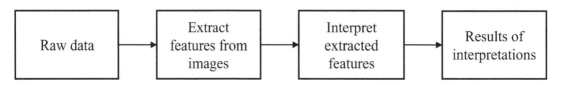

Figure 2-14. *Conceptual components of an image-based pretrained model*

Feature extraction serves the purpose of assembling information into meaningful observations, like identifying and amplifying (or reducing) the presence of certain edges, shapes, or color patterns. A well-tuned feature extraction component is able to identify and amplify characteristics that are relevant to the problem. For instance, if the problem is to classify various grayscale images of shapes, the feature extraction component must be highly capable at detecting and amplifying the shape of edges.

Feature interpretation interprets the compiled extracted features to make a final judgment about the image. It takes in extracted information regarding the features of the image and can perform comparisons and other complex analyses across various regions of the image. A well-tuned feature interpretation component is able to effectively aggregate and make sense of extracted features in relation to the target output.

In practice, the roles of "feature extraction" and "feature interpretation" are played by convolutional and fully connected components (Figure 2-15). The convolutional component generally consists of layers like convolutions, pooling, and other image-based feature extraction layers. The fully connected component generally consists of fully connected or dense (in Keras terminology) layers.

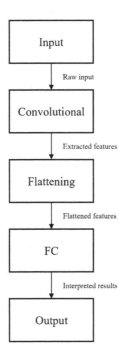

Figure 2-15. *The practical components of an image-based pretrained model*

1. The raw input is passed into the convolutional component for feature extraction.

2. The convolutional component of the pretrained model takes in the raw input and performs convolutions, pooling, and other image-processing layers to extract features from the image. It then passes the extracted features into the flattening component.

3. The flattening component takes in the two-dimensional (spatially, not counting color depth), since the convolutional component performs operations that take in and output data of image form. Flattening converts the two-dimensional extracted features into one-dimensional data so that the fully connected layer can operate upon the features.

4. The fully connected component, also known as the "top" or
    the "head" of the neural network, takes in the one-dimensional
    flattened data containing the information from the extracted
    features. After interpreting the extracted features through a series
    of fully connected layers (and other corresponding layers, like
    activations, dropout, etc.), the fully connected layer outputs the
    output of the neural network.

We'll talk more about certain architectures in depth in later chapters, but here we'll
do a very quick overview of important models.

In Keras, pretrained models are arranged into modules from which that model
and related functions can be found within `keras.applications.module_name`. For
instance, if you wanted to find the pretrained model and processing functions relating
to the InceptionV3 model, you would find it in `keras.applications.inception_v3`.
In addition to the pretrained model object, these modules often contain processing
functions to apply to your data or the model output. We will see how these processing
functions work in relation to the model soon.

The ImageNet Dataset

ImageNet is one of the most important datasets in image recognition. Professor Fei-Fei
Li began working on ImageNet at Princeton in early 2007. Throughout the development
of neural network applications in the image domain, ImageNet has been a core dataset
upon which new methods and designs were conceived, tested, and reinvented.

The ImageNet dataset is structurally organized according to the WordNet hierarchy,
in which objects are arranged in hierarchical relation to each other (Figure 2-16).

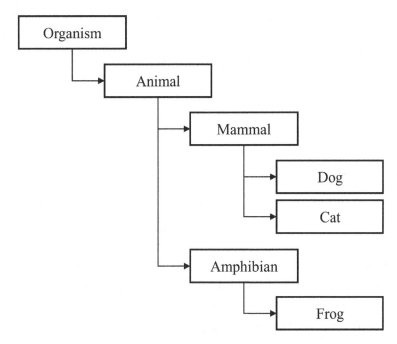

Figure 2-16. *Hypothetical example branch of WordNet hierarchy organization*

Each "meaningful concept" in ImageNet is referred to as a "synonym set" or a "synset." Each "synset" can be described by multiple words or series of words. ImageNet contains over 100,000 synsets; each synset contains an average of 1000 images. Across 22,000 objects, ImageNet contains over 14 million images.

The ImageNet project was inspired by two important needs in computer vision research. The first was the need to establish a clear North Star problem in computer vision... Second, there was a critical need for more data to enable more generalizable machine learning methods. Ever since the birth of the digital era and the availability of web-scale data exchanges, researchers in these fields have been working hard to design more and more sophisticated algorithms to index, retrieve, organize and annotate multimedia data. But good research requires good resources...The convergence of these two intellectual reasons motivated us to build ImageNet.

—ImageNet Research Team[5]

---

[5] https://image-net.org/about.php.

The ImageNet Large Scale Visual Recognition Challenge ran from 2010 to 2017 and was instrumental to the development of image-processing deep learning work. Hence, many of the pretrained models available in Keras/TensorFlow and other deep learning frameworks are pretrained on the ImageNet dataset. Models pretrained on the ImageNet dataset have developed skills to recognize edges, textures, and other attributes of images of three-dimensional spaces and objects.

ResNet

The ResNet architectures, or residual neural networks, were introduced in 2015. ResNet won first place in the 2015 ImageNet challenge, obtaining a 3.57% error on the test dataset. The ResNet uses residual connections, or skip connections, which are connections between layers that skip over at least one intermediate layer. We'll explore these sorts of designs and strategies in neural network architectures in later chapters.

Keras/TensorFlow offers several different versions of the ResNet architecture: ResNet50, ResNet101, ResNet152, ResNet50V2, ResNet101V2, and ResNet152V2. The number after the "ResNet" (e.g., the "50" in "ResNet50") indicates the number of layers in it. Thus, ResNet152V2 has more layers than ResNet101V2.

ResNet also comes in two versions, with two (very general) key differences:

- ResNetV1 adds more nonlinearities than ResNetV2. The V2 architecture thus allows for a more direct and clear path for data flow throughout the network.

- ResNetV1 passes the data through a convolutional layer before batch normalization and an activation layer, whereas ResNetV2 passes data through the batch normalization and activation layers before the convolutional layer (reversed).

You can find Keras ResNet model architectures in corresponding modules within `keras.applications`:

- ResNet50: `keras.applications.resnet50.ResNet50()`

- ResNet101: `keras.applications.resnet.ResNet101()`

- ResNet152: `keras.applications.resnet50.ResNet152()`

- ResNet50V2: `keras.applications.resnet_v2.ResNet50V2()`

- ResNet101V2: `keras.applications.resnet_v2.ResNet101V2()`

- ResNet152V2: `keras.applications.resnet_v2.ResNet152V2()`

We will explore how to use the architecture given the model object later.

InceptionV3

The 2015 InceptionV3 model is a popular model from the Inception family of models, using a module/cell-based structure in which certain sets of layers are repeated. Inception networks are more computationally efficient by reducing the number of parameters necessary and limiting the memory and resources needed to be consumed. InceptionV3 was designed primarily with a focus on minimizing computational cost, whereas ResNet focuses on maximizing accuracy.

You can find the InceptionV3 model architecture at `keras.applications` `.inception_v3.InceptionV3()`.

Inspired by the high performance of ResNet, the Inception-ResNet architecture uses a hybrid module by incorporating residuals. Correspondingly, Inception-ResNet generally has a low computational cost with high performance.

You can find the Inception-ResNetV2 architecture at `keras.applications` `.inception_resnet_v2.InceptionVResNet2()`.

For more specific and technical discussion on residual connections, cell-based structures, and the InceptionV3 architecture, see Chapter 6.

MobileNet

The 2017 MobileNet models are designed to perform well on mobile phone deep learning applications. MobileNets use depth-wise separable convolutions, which are convolutions that apply not only spatially but also depth-wise. MobileNet has more parameters than Inception but less than ResNet; correspondingly, MobileNet has been generally observed to perform worse than ResNet but better than Inception.

Note   When we compare models (e.g., "MobileNet vs. ResNet"), keep in mind that we are referring to the complete family of architectures. Most models have architectures of different versions and depths. When comparing model families, we are referring to architectures of comparable versions and layers.

Keras/TensorFlow offers four versions of MobileNet, MobileNetV1, MobileNetV2, MobileNetV3Small, and MobileNetV3Large. MobileNetV2, like MobileNetV1, uses depth-wise separable convolutions but also introduces linear bottlenecks and shortcut connections between bottlenecks. MobileNetV2 can effectively extract features for object detection and segmentation and generally performs faster at achieving the same performance as MobileNetV1. MobileNetV3Small and MobileNetV3Large are MobileNet architectures designated for low-resource and high-resource consumption scenarios, derived from Neural Architecture Search algorithms (we will discuss Neural Architecture Search and other methods in later chapters).

You can find Keras ResNet model architectures in corresponding modules within `keras.applications`:

- MobileNetV1: `keras.applications.mobilenet.MobileNet()`

- MobileNetV2: `keras.applications.mobilenet_v2.MobileNetV2()`

- MobileNetV3Small: `keras.applications.MobileNetV3Small()`

- MobileNetV3Large: `keras.applications.MobileNetV3Large()`

EfficientNet

Deep convolutional neural networks have continually been growing larger in an attempt to be more powerful. Exactly how this enlargement is performed, however, has varied. Some approaches increase the resolution of the image by increasing the number of pixels in the image handled by the network. Others increase the depth of the network by adding more layers or the width by increasing the number of nodes in each layer.

The 2019 EfficientNet family of models is a new approach for neural network enlargement via compound scaling, in which the resolution, depth, and width of the network are equally scaled. By using this scaling method upon a small, base model named EfficientNetB0, seven compound-scaled architectures were generated, named EfficientNetB1, EfficientNetB2, …, EfficientNetB7 as the magnitude of scaling increases. The EfficientNet family of models is both powerful and efficient, both computationally and time-wise. Compound scaling allows EfficientNet to improve upon the performance of models like MobileNet and ResNet on the ImageNet dataset.

You can find the Keras implementation of EfficientNet model architectures in the corresponding module within `keras.applications`:

- EfficientNetB0: `keras.applications.efficientnet.EfficientNetB0()`

- EfficientNetB1: `keras.applications.efficientnet.EfficientNetB1()`

- EfficientNetB2: `keras.applications.efficientnet.EfficientNetB2()`

  …

- EfficientNetB6: `keras.applications.efficientnet.EfficientNetB6()`

- EfficientNetB7: `keras.applications.efficientnet.EfficientNetB7()`

Find a more specific and technical discussion of EfficientNet in a Chapter 6 case study.

Other Models

Keras/TensorFlow offers a host of other pretrained models in `keras.applications`. To view them, refer to the TensorFlow documentation, which not only provides information on usage, parameters, and methods but also includes the links to each pretrained model's corresponding paper: `www.tensorflow.org/api_docs/python/tf/keras/applications`. Later in this chapter, we will also cover how to convert PyTorch models into Keras/TensorFlow models to take advantage of PyTorch's library of pretrained models.

Changing Pretrained Model Architectures

To change pretrained model architectures, usually only the pretrained model's convolutional component is transferred. Recall that transfer learning is primarily concerned with transferring the skills learned from one general dataset for application in a specific task dataset. Because the task dataset is different from the general dataset, while a model operating on the task dataset would still benefit from the feature-extracting skills transferred via transfer learning, it would need to develop its own interpretations.

As a conceptual analogy, imagine you are an art student trained in the analysis of Impressionist-era works – the art of Monet, Renoir, Matisse, and so on. If now you were to approach a contemporary work by the likes of Picasso and Pollock, you would want to

- *...retain the feature-extracting skills*: The fundamental feature extraction capabilities that you learned as an art student are important to developing and organizing meaningful and astute observations that can be used for interpretation. Without understanding how to look at and take away observations of art (without actually interpreting them), there are no observations to serve as the basis of interpretation.

- *...but develop a fresh set of interpretations*: Now that you have kept your feature extraction skills and have developed a set of observations (the extracted features), it wouldn't in most contexts make sense to analyze the contemporary work through Impressionist interpretations. Rather, this new data calls for a fresh set of interpretations suited toward the context of contemporary art.

Thus, while the skills for extracting features are valuable and can be transferred almost universally across most generally similar problems, often the interpretations of said features vary widely enough that the most successful strategy is to build a custom fully connected component, instead of transferring it (Figure 2-17).

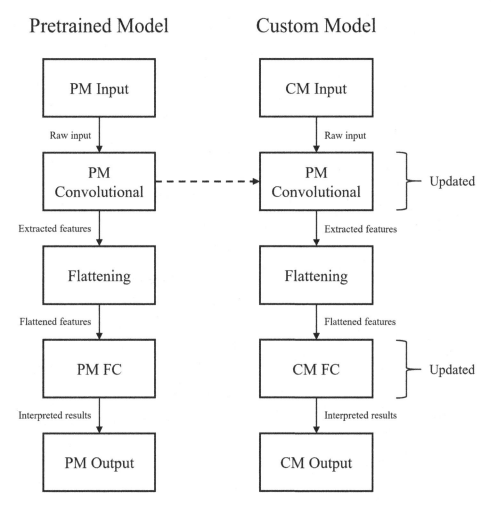

Figure 2-17. *The process of transferring weights from a pretrained model to a custom model. PM = pretrained model, CM = custom model. "Updated" indicates that the corresponding component is being trained/updated in retraining*

However, in some occasions the problem set may be similar enough that the fully connected layer weights may also be transferred (Figure 2-18). This method requires no architecture manipulation of the pretrained model, since all the weights from the pretrained model are being transferred. To use this method, simply instantiate the pretrained model and train it on the task dataset without any architecture manipulation.

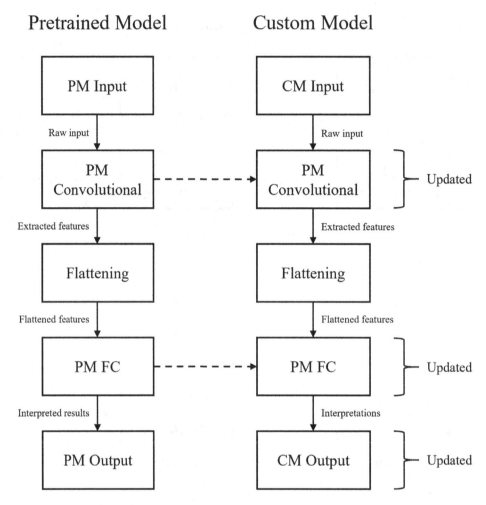

Figure 2-18. *An alternative process of transferring weights from a pretrained model to a custom model in which both the pretrained model's convolutional and FC components are transferred*

Another alternative is to keep the pretrained model's fully connected component and to add more custom layers afterward (Figure 2-19). Think back to the art student analogy: you're starting from the interpretative perspective of the Impressionist style that you were trained in, but additionally developing new interpretations *of* your interpretations to account for the new problem type. However, this method is generally less often used than only transferring weights in the convolutional component, since to be effective the neural network must be significantly deepened (more layers added), which can pose problems for updating weights, time, and model performance. Under certain conditions, though, it can be the optimal strategy.

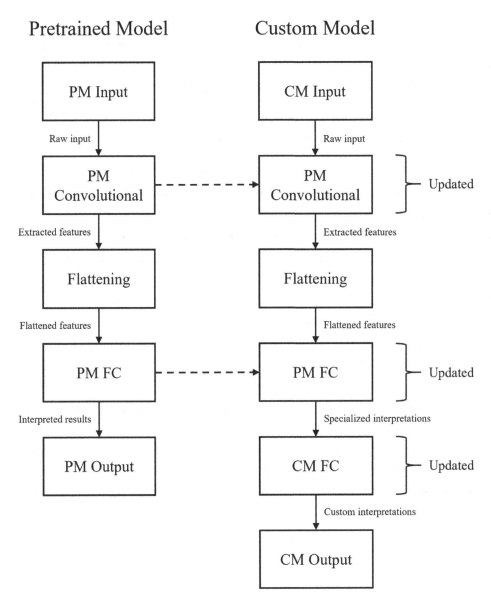

Figure 2-19. *An alternative process of transferring weights from a pretrained model to a custom model in which both the pretrained model's convolutional and FC components are transferred, and new fully connected layers are added to the custom model*

Although we will only explore the first method through examples, all can be implemented through tools covered in Chapter 1, like the Functional API. Models can be treated like functional layers, expressed as functions of an input. See Chapter 3 for more complex exploration into neural network architecture manipulation.

Neural Network "Top" Inclusivity

All Keras/TensorFlow pretrained models have an important parameter, named `include_top`, which allows us to easily implement the first method of changing pretrained model architectures for training.

Recall that the fully connected component of an image-based neural network architecture is often referred to as its "top." Thus, if the `include_top` parameter is set to `True`, the fully connected component will be retained in the instantiated pretrained model object. On the other hand, if the `include_top` parameter is set to `False`, the fully connected component will not be retained and the instantiated pretrained model object will contain only the input and convolutional components.

It's important to realize that when there is no top, the input image shape is more or less arbitrary (see Note) because convolutional layers do not rely upon an absolute shape. A convolutional layer can act on an image of almost any shape (see Note for the qualifier "almost"), since it simply "slides" filters over an image. On the other hand, a dense layer in the fully connected component can only act upon inputs of a certain shape. Thus, if the `include_top` parameter is set to `True`, you are also limited in the shape of your input data to whatever the shape of the data the model was pretrained on.

Note   The input image shape is not completely arbitrary. If the image shape is too small, the image may not be large enough for the whole depth of the network to be able to be valid, since filters rely on a minimum image size. One cannot perform ten convolutions with kernel size (3,3) on a 16x16 pixel image because in the middle of the sequence the feature map will have been reduced so much that it is too small to perform convolutions on. This is likely the culprit if you are using a model architecture with a small input size that throws an error when defining the architecture.

It should also be noted that the fact that the absence of a top means the input size is somewhat arbitrary does not mean that you should be too extreme with how much you change the input shape of your image. Say, a model was trained on 512x512 pixel images and you used transfer learning by extracting the convolutional component of that model (along with its weights). If you were to train the model on 32x32 pixel images (likely with image resizing), even though the operation is technically valid, the transferred weights are rendered more or less useless because the skills in the original model were developed on more high-resolution pixel images. In short, make use of the relatively arbitrariness of image sizes with no top, but also take your freedom in image size with a grain of salt.

Layer Freezing

Layer freezing is a useful tool for the implementation of transfer learning. When a layer is "frozen," its weights are fixed and it cannot be trained. The purpose of transfer learning is to utilize important skills the pretrained model has developed. Thus, it is common to freeze the convolutional component of the pretrained model and to train only the custom fully connected component (Figure 2-20). This way, the feature-extracting skills learned and stored in the weights of the convolutional component are kept as is, but the interpretative fully connected component is trained to best interpret the extracted features. After this, sometimes the entire network is trained for fine-tuning. This allows the convolutional component to be updated in accordance with the developed interpretations in the fully connected component. Be aware that too much fine-tuning can lead to overfitting.

Think of layer freezing in relation to the art student analogy. Initially, you want to utilize the feature-extracting skills you developed in your art training and to construct new interpretations of contemporary art corresponding to the extracted features. However, after you have constructed new interpretations for the particular problem, you may find it helpful to go back and slightly update your feature-extracting skills to better service your new interpretative capabilities.

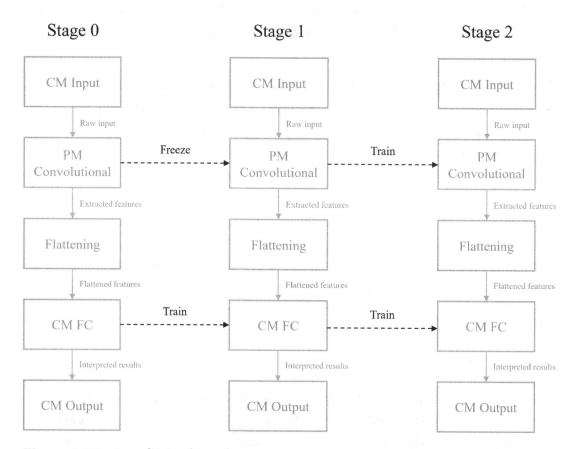

Figure 2-20. *A multistep layer freezing process*

On a practical level, layer freezing is immensely helpful for practical success. Although it can be easy to build massive neural networks and call .fit() with increasing computational power, it remains true that it is not easy to optimize hundreds of millions (or even billions) of parameters by brute-force and achieve good performance. Layer freezing allows the neural network to focus on optimizing one segment of its architectures at a time and allows you to better use the weights gained from pretraining by restricting how much the neural network can deviate from the learned pretrained weights.

Implementing Transfer Learning

In this section, we will discuss three examples of implementing transfer learning with Keras/TensorFlow:

1. *No architecture or weight changes*: An exercise in using (ImageNet) pretrained models for their original purpose – to predict what object is in an image. Also helpful for dealing with reshaping inputs and using encoding and decoding functions associated with the pretrained model module

2. *Transfer learning without layer freezing*: Using the standard procedure of transferring the weights from the convolutional component and building a custom fully connected component, without layer freezing. Helpful for manipulating model architectures both using the Functional API and using the `include_top` parameter

3. *Transfer learning with layer freezing*: Using the standard procedure of transferring the weights from the convolutional component and building a custom fully connected component, with layer freezing

We'll also discuss how to take advantage of PyTorch's pretrained model library by converting them into Keras/TensorFlow models.

General Implementation Structure: A Template

Before we build a transfer learning model, let's first compile the topics and methods discussed before to lay out a template for the general implementation structure of transfer learning:

1. Choose an appropriate model from the Keras/TensorFlow repository of pretrained models or from another source, like a pypi library offering an implementation of a pretrained model or PyTorch.

2. Instantiate the model with the desired architectural settings (include or do not include top).

3. Set up the model's input and output flows such that it matches
    with the data. This may require using a reshaping layer, using an
    input layer, adding a fully connected component such that the
    extracted features are interpreted and outputted, some other
    mechanism, or some combination.

4. Freeze layers, as necessary.

5. Compile and fit.

6. Change which layers are frozen if necessary, and compile and fit
    again.

It should be also noted that although in this context we are discussing image-based
pretrained models, as transfer learning began its extensive development in the domain
of images, the logic of a feature-extracting segment and an interpretative segment can be
applied to other problem domains as well.

No Architecture or Weight Changes

If we make no changes to the architecture, we cannot adjust the input or the output of
the pretrained model. Therefore, if we want to make no changes to the architecture, we
must use the pretrained model for its original purpose.

The InceptionV3 model was trained on the ImageNet dataset. Thus, it takes an image
of size $(299,299,3)$ and returns an output vector with the probability the image belongs
to one of 1000 classes.

Let's begin by loading an example image to demonstrate how InceptionV3 would
function (Listing 2-1). We will use requests to read a URL, PIL to read the corresponding
image (Figure 2-21), and numpy to convert the image into an array.

Listing 2-1. Loading image of a lion

```
import PIL, requests, numpy as np
from PIL import Image
url = 'https://cdn.pixabay.com/photo/2018/04/13/21/24/
lion-3317670_1280.jpg'
im = np.array(Image.open(requests.get(url, stream=True).raw))
```

Figure 2-21. *Sample image to be classified by a model architecture trained on the ImageNet dataset*

Let's begin by importing necessary layers, applications, and functions (Listing 2-2):

- The keras.layers.Input layer defines the input size and is required for constructing a neural network (or in some other form, like the input_shape parameter).

- The keras.layers.experimental.preprocessing.Resizing layer resizes an image to a new shape. We can use this to reshape an image of any size to the shape (299,299,3) such that it is in the proper input shape for InceptionV3.

- The keras.applications.inception_v3.InceptionV3 model is the core InceptionV3 model that can be used and manipulated in relation to the other layers.

- The keras.applications.inception_v3.preprocess_input preprocesses data to be in the same format InceptionV3 was trained on. If the inputs are not preprocessed, the pretrained weights may not be suited toward the unprocessed inputs and yield inaccurate results.

Listing 2-2. Importing necessary libraries

```
from keras.layers import Input
from keras.layers.experimental.preprocessing import Resizing
from keras.applications.inception_v3 import InceptionV3
from keras.applications.inception_v3 import preprocess_input
```

We can use these layers to set up a series of data input and preprocessing (Listing 2-3). First, the Input layer takes in data with a shape of (None, None, 3); None indicates that the exact value for that dimension is not exactly specified. This allows us to pass images of any size into the neural network input. The Resizing layer reshapes the input image to the proper shape for InceptionV3. The third layer preprocesses the resized data into InceptionV3 format.

Listing 2-3. Building input of a transfer learning neural network

```
input_layer = Input((None, None, 3))
reshape_layer = Resizing(299,299)(input_layer)
preprocess_layer = preprocess_input(reshape_layer)
```

The output of preprocess_layer (the result of the resized, preprocessed input data) is passed as the input to the InceptionV3 model. Note that we are making no changes to the model architecture or weights, so we set inlucde_top=True and weights='imagenet'. We can treat the model as a layer that takes in the output of a previous layer and can be passed as an input to the following layer: Inceptionv3 = InceptionV3(include_top=True, weights='imagenet')(preprocess_layer).

Note that you can set weights to None to just use the model architectures, without the pretrained weights. Although this doesn't quite count as transfer learning, you may want to use this when the task dataset is so different from the pretraining dataset that any pretrained weights wouldn't be of much benefit, but want to take advantage of some architecture's characteristics, like efficiency or power.

We can create a model out of this set of layers using keras.models.Model (note that the pretrained model is treated like a layer and thus is considered an outputs layer): model = keras.models.Model(inputs=input_layer, outputs=Inceptionv3).

Now that the model has been created, we can use it to make predictions (Listing 2-4). We will use the decode_predictions function to help us parse the outputs of the InceptionV3 model, which in its raw form is a list of numbers. The model predictions for the image (stored in the variable im) can be passed into decode_predictions to obtain interpretable results.

Listing 2-4. Running and decoding predictions for a pretrained model

```
from keras.applications.inception_v3 import decode_predictions
reshaped_im = np.array([im])
print(decode_predictions(model.predict(reshaped_im)))
```

Note that the model expects four-dimensional data – for example, data with shape $(100,299,299,3)$, indicating that it consists of 100 299x299 pixel RGB colored images. Even though we are submitting an individual image for prediction, it still needs to be four-dimensional. An easy way to accomplish this is to wrap it as an element in another array with `np.array([im])`.

Each element of the results of decoded predictions is in the format (`class name, class description, probability`). The results of the decoded predictions show that InceptionV3 has correctly identified the object of the image – a lion, with 91% confidence (Listing 2-5).

Listing 2-5. Results of InceptionV3 decoded predictions on an image of a lion

```
[[('n02129165', 'lion', 0.91112673),
  ('n02112137', 'chow', 0.008575964),
  ('n02130308', 'cheetah', 0.0024228022),
  ('n04328186', 'stopwatch', 0.00097137265),
  ('n02106030', 'collie', 0.00083191396)]]
```

A similar process can be applied to other pretrained models available via Keras. Most Keras pretrained models have associated `preprocess_input` and `decode_predictions` functions.

Transfer Learning Without Layer Freezing

If we want to adapt the pretrained model for our own task dataset, we need to make some minimal architecture changes such that the input and the output can accommodate our dataset. This follows a very similar structure to the previous application of no architecture or weight changes.

Let's begin by importing necessary layers and models (Listing 2-6):

- The `keras.layers.Input` layer defines the input shape.

- The `keras.layers.Dense` layer provides the neural network's "interpretative" or "predictive" power.

- The `keras.layers.GlobalAveragePooling2D` layer "collapses" image data into the average of its elements. For instance, image data with shape (`a, b, c, d`) will have shape (`a, d`) afterward, where (`b, c, d`) is the shape of each image element and there are `a` training instances. An alternative to Global Average Pooling is the raw Flatten layer. The Flatten layer (`keras.layers.Flatten()`), unlike the Global Average Pooling layer, retains all the elements of multidimensional data and converts them into one-dimensional data by stacking them end to end. Thus, image data with shape (`a, b, c, d`) will have shape (`a, b*c*d`).

- The `keras.applications.inception_v3.InceptionV3` model provides the architecture and pretrained weights for the InceptionV3 model.

Listing 2-6. Importing important layers for transfer learning

```
from keras.layers import Input, Dense, GlobalAveragePooling2D
from keras.applications.inception_v3 import InceptionV3
```

Let's begin by defining the input layer and the InceptionV3 model it feeds into (Listing 2-7).

Listing 2-7. Building the input and pretrained model of a transfer learning model

```
input_layer = Input((512,512,3))
inceptionv3 = InceptionV3(include_top=False,
                          weights='imagenet')(input_layer)
```

Note that we are keeping the ImageNet weights, but we do not include the top. While we want to keep the feature extraction skills in the convolutional layers, because we are appropriating the pretrained model for our own purposes, we don't need the weights for the interpretative fully connected component.

Because we are not keeping the top of the neural network, the output of the truncated InceptionV3 model is an image. The Global Average Pooling model allows us to compress the data out of image form (this is the flattening component). Afterward, we can place two more dense layers to "interpret" the extracted features and one more to form the output layer (Listing 2-8).

Listing 2-8. Building the flatten and custom FC components of the transfer learning model

```
pooling = GlobalAveragePooling2D()(inceptionv3)
custom_dense_1 = Dense(256, activation='relu')(pooling)
custom_dense_2 = Dense(64, activation='relu')(custom_dense_1)
output_layer = Dense(1, activation='sigmoid')(custom_dense_2)
```

We can form a model as such (Listing 2-9).

Listing 2-9. Compiling the transfer learning model into a Keras model

```
model = keras.models.Model(inputs = input_layer, outputs = output_layer)
```

Visualizing the model shows the shape of the data over time (Figure 2-22) with keras.utils.plot_model(model, show_shapes=True). The visualization also demonstrates that the InceptionV3 model is treated very much like a standard layer.

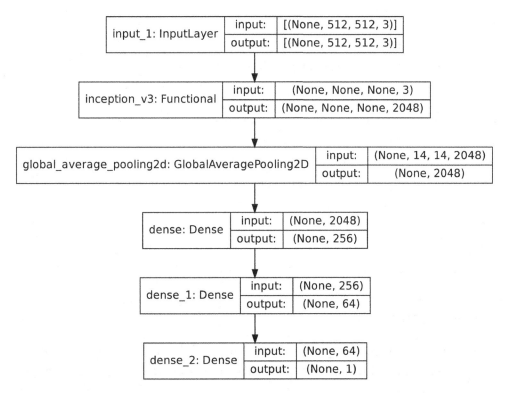

Figure 2-22. *Diagram of an example model architecture using transfer learning*

The model can be correspondingly compiled and trained. It should be noted that usually the fully connected component is much longer and complex to account for the complex nature of modern deep learning datasets.

Transfer Learning with Layer Freezing

In order to freeze a layer, set `layer_obj.trainable=False`.

We need to first instantiate the layer to set the `trainable` attribute to `False` (Listing 2-10).

Listing 2-10. Instantiating the pretrained model as a layer without directly taking in inputs

```
inception_model = InceptionV3(include_top = False, weights = 'imagenet')
```

We can then use this pretrained model object as a standard layer in conjunction with other layers using the syntax of the Functional API (Listing 2-11).

Listing 2-11. Compiling layers into a transfer learning model

```
input_layer = Input((512,512,3))
inceptionv3 = inception_model(input_layer)
pooling = GlobalAveragePooling2D()(inceptionv3)
...
```

After building the remainder of the model and aggregating the layers into a model with `keras.models.Model`, you can call `model.summary()` to get a rundown of the model's parameters (Listing 2-12). Note that most of the parameters are trainable. A small quantity of non-trainable parameters are built into the model's architecture.

Listing 2-12. Number of trainable and non-trainable parameters before layer freezing

```
Total params: 22,343,841
Trainable params: 22,309,409
Non-trainable params: 34,432
```

You can now call `inception_model.trainable = False` to freeze the layer. By re-aggregating the layers into a new model and calling `model.summary()`, you can see the results of layer freezing (Listing 2-13). Because most of the parameters in the complete model were transferred weights from the pretrained model, after layer freezing, most of the parameters are non-trainable.

Listing 2-13. Number of trainable and non-trainable parameters after layer freezing

```
Total params: 22,343,841
Trainable params: 541,057
Non-trainable params: 21,802,784
```

Calling `model.summary()` is a good way to double-check if layer freezing worked or not. When several forms and series of neural network manipulations are being performed, it can be difficult to keep track and separate different forms of functions and instantiated objects.

You can unfreeze with `inception_model.trainable = True` and continue fitting again for fine-tuning.

You can also freeze individual layers with a similar method or by aggregating layers together into models and unfreezing groups of layers.

Accessing PyTorch Models

Keras/TensorFlow offers a large selection of pretrained models, and platforms like the model zoo (`https://modelzoo.co/`) or pypi libraries offer a wide range of pretrained models and/or architectures in Keras/TensorFlow.

PyTorch, a major alternative framework to Keras/TensorFlow, offers a larger selection of pretrained models. If a PyTorch pretrained model is not already implemented in Keras/ TensorFlow, there's a simple method to convert PyTorch models into Keras model objects so you can work with the model using familiar methods and steps.

PyTorch and Keras/TensorFlow, as well as most other frameworks, are built upon the Open Neural Network Exchange (ONNX), an AI ecosystem that establishes standards for representing various AI algorithms. ONNX was formulated with the purpose of encouraging innovation via conversion of algorithm representations across frameworks. Correspondingly, to convert a model from PyTorch to Keras/TensorFlow, the PyTorch model is first converted into ONNX format and then from ONNX format to the Keras/ TensorFlow framework (Figure 2-23).

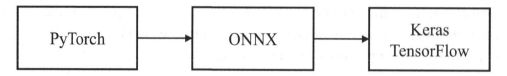

Figure 2-23. *Diagram converting a PyTorch model to a Keras/TensorFlow model*

Let's start by installing PyTorch with `pip install torch`. It can be imported as `import torch`.

You can find a list of PyTorch image-based pretrained models at `https://pytorch .org/vision/stable/models.html` (note that you can access other non-image-based pretrained models as well). These image-based pretrained models can be accessed at `torchvision.models.model_name`. You can install TorchVision via pip with `pip install torchvision`.

For the purposes of this example, we will convert the PyTorch implementation of the SqueezeNet model into Keras. The 2016 SqueezeNet model focuses on retaining model performance while decreasing computational cost, notably by using 1x1 convolutions to reduce the number of parameters. The first step is to instantiate the PyTorch model (Listing 2-14).

Listing 2-14. Using PyTorch to instantiate a PyTorch pretrained model

```
from torchvision import models
pytorch_model = models.squeezenet1_0(pretrained=True)
```

Now that we have instantiated the PyTorch model, we can convert it into a Keras model using the `pytorch2keras` model, which can be installed via pip with `pip install pytorch2keras`.

Let's begin by importing the `pytorch_to_keras` function (Listing 2-15).

Listing 2-15. Importing the necessary function to convert a PyTorch model to a Keras model

```
from pytorch2keras.converter import import pytorch_to_keras
```

The `pytorch_to_keras` function takes in three arguments: the PyTorch model, the input tensor, and the shape of the input tensor (Listing 2-16).

Listing 2-16. Converting a PyTorch model to a Keras model. Note that we did not explicitly import it, but np refers to numpy; this can be imported as import numpy as np

```
from torch.autograd import Variable
from torch import FloatTensor
input_np = np.random.uniform(0, 1, (1, 3, 256, 256))
input_var = Variable(FloatTensor(input_np))
keras_model = pytorch_to_keras(pytorch_model,
                               input_var,
                               [(3, None, None,)])
```

The input_np variable defines the shape of the input in channels-first format, where the "depth" of the image is listed as the first element of the shape. The torch. FloatTensor and torch.autograd.Variable functions convert the shape of the numpy array into valid tensor format.

The resulting Keras model can be visualized to double-check that the architecture has been transferred correctly and then compiled and fitted.

Note that not all PyTorch architectures can be converted into Keras/TensorFlow if some layer or operation is not supported for export by ONNX. See a list of transferable layers on the pytorch2keras GitHub: https://github.com/gmalivenko/pytorch2keras.

Implementing Simple Self-Supervised Learning

In this section, we will discuss the implementation of rudimentary level of self-supervised learning. You will find that many deep learning concepts are intertwined, and thus we need to wait until Chapter 3 because we first need to have a strong grasp of what autoencoders are before we understand their effective use in pretraining.

In more complex forms of self-supervised learning, modifications to the architecture are performed after the pretraining task often by adding more layers, usually to get the output into the desired shape to train on the task dataset or to add more processing power. In a simpler application of self-supervised learning, we will design our self-supervised pretraining strategy such that no architectural changes need to be made after pretraining and before formal training on the task dataset.

In order to design a self-supervised pretraining strategy that does not require any architecture changes, we need to focus primarily on the input and the output. If the shapes of the pretraining task's x and y match that of the ultimate task's, then there is no need to alter the architecture to accommodate it.

Consider the task of classifying the gender of a person in an image. Here, we have x = image and y = 1-dimensional vector (one value: 0 or 1, indicating gender). Thus, our self-supervised pretraining task should similarly require the model to take in an image and output one value.

For this example, the self-supervised pretraining task will be to identify the degree of noise. For each training example, we will add Gaussian noise with a certain standard deviation, σ. The objective of the model for pretraining is to predict the degree of noise, σ, given the noisy image. In order to perform such a task, the model would need to identify which features in the model count as noise and which ones do not, and thus it works toward discovering and representing fundamental structures and patterns in faces. Adding Gaussian noise changes the color of the images, not the "location" of certain features; since color determines the presence of shadows, edges, and other characteristics that define what a female and male face would generally look like, though, a good understanding of common relationships between the color of pixels between images would be valuable. Moreover, it moves the model closer toward being able to be robust to small perturbations and noise.

Fundamentally, at a rudimentary level without architecture changes, building self-supervised learning pretraining strategies is concerned primarily with data. We can build two datasets – augmented task dataset used for pretraining and a task dataset used for formal training – and fit the model on those two datasets, in that order. After being trained on the augmented task dataset, ideally the model will have developed representations of the "world" it is about to enter and can approach the ultimate task more aware and with more complex understandings than if it had not undergone pretraining.

To construct the augmented task dataset, we will use the `.map` method of loading images from directories into the efficient TensorFlow dataset format (see Chapter 1). For brevity, here we will provide only the function to be mapped in constructing the altered dataset.

Let us first define a function, `parse_file`, which takes in the path of a file and uses TensorFlow input/output and image operations to read the image and convert it into proper format (Listing 2-17). The function returns a tensor containing the information from the image, upon which we can perform operations.

Listing 2-17. Function to parse a file path using TensorFlow functions

```
def parse_file(filename):
    raw_image = tf.io.read_file(filename)
    image = tf.image.decode_png(raw_image, channels=3)
    image = tf.image.convert_image_dtype(image, tf.float32)
    image = tf.image.resize(image, [512,512])
    return image
```

The alter_data function will be the "ultimate function" used in .map() (Listing 2-18). First, we use the parse_file function to collect information from the image. Next, we will use TensorFlow's tf.random.uniform() function to randomly choose a number from a uniform distribution ranging from 0 to 0.5; this will be the standard deviation of the Gaussian noise applied. Using the standard deviation and mean 0.0, we will produce a Gaussian noise tensor with the same shape as the image. The noise can be applied to the image by adding the image tensor and the noise tensor together. Lastly, we return the altered image (which now has noise applied to it) and the standard deviation (the label). The task dataset has been altered such that the pretrained task is to predict the degree of noise applied to the image.

Listing 2-18. TensorFlow function to implement an alteration to the dataset for pretraining

```
def alter_data(filename, label):
    image = parse_file(filename)
    std = tf.random.uniform(shape=[1], minval=0, maxval=0.5)
    noise = tf.random.normal(shape=tf.shape(image),
    mean=0.0, stddev=std, dtype=tf.float32)
    image = tf.add(image, noise)
    return image, std
```

If you had a multidimensional output, you could also have the model predict changes to brightness, hue, etc. – features that are important to this task. TensorFlow offers a host of image processing and mathematical operations that you can use to alter data. It is best to stick to TensorFlow functions and objects when constructing functions to map to avoid errors or computational inefficiency.

We can see the visual results of our alteration (Figure 2-24).

Figure 2-24. *Example images from the altered task dataset. The standard deviations of noise (labels) for images from left to right are 0.003, 0.186, and 0.304*

You can then fit the model (after compiling and performing other necessary steps) on the two datasets (Listing 2-19). Note that the number of epochs the model uses fitting on the altered task dataset compared to the number of epochs used fitting on the task dataset varies significantly by the complexity of the task and the design of the altered task dataset, as well as other factors like computational resources available and the size of the datasets. If the task is immensely complex and the altered task dataset offers difficult to obtain but meaningful and deep representations of the task dataset (for instance, modern NLP problems), it may be fruitful to train the model on the altered task dataset for a significant period of time. On the other hand, in simpler tasks, like in this example of classifying the gender of a person in an image, and with altered task datasets that offer meaningful representations but are not integral toward representing the world of the task dataset, pretraining probably does not need to last as long. Getting a feel for the relationship between pretraining and formal training takes experience and time, but it's good to experiment with your unique conditions and environment.

Listing 2-19. Performing pretraining and ultimate training

```
model.fit(altered_task_dataset, epochs=10)
model.fit(task_dataset, epochs=40)
```

It should also be noted that the altered task dataset and the task dataset in this case are different types of problems – the former is a regression problem (predicting a continuous value), whereas the latter is a classification problem (categorizing an image). In this case, this difference is not a problem because all labels in both datasets reside between 0 and 1, meaning that a standard model with a sigmoid activation function output can operate on both. Binary cross-entropy is also a valid loss function for regression problems bounded between 0 and 1, although if you would like you can compile the model with a formally regression-based loss function (e.g., MSE, Huber) before training on the altered task dataset and recompile the model with binary cross-entropy (or some other loss function for classification) before training on the task dataset. In more advanced cases, adjusting from the altered task dataset to the task dataset requires making architectural modifications. You'll find examples of this in the second case study of this chapter, as well as in Chapter 3 on autoencoders. With the tools of autoencoders, we can perform more complex pretraining strategies, like using image-to-image pretraining tasks (e.g., colorization, denoising, resolution recovery) to pretrain a model to perform an image-to-vector or image-to-image task.

Case Studies

These three case studies allow us to explore new ideas and recent research in transfer learning and self-supervised learning. Some ideas may be out of the scope of this book in terms of implementation but offer fresh perspectives and things to think about when designing your deep learning approaches.

Transfer Learning Case Study: Adversarial Exploitation of Transfer Learning

As deep learning models are increasingly deployed in common production, it is becoming more important to incorporate the findings from the rising field of adversarial learning into deep learning designs. Adversarial learning is concerned with how to exploit weaknesses in deep learning models, often by making small changes that are imperceptible to the human eye but that completely change the model's output. Vulnerabilities in deep learning can lead to dangerous outcomes – imagine, for instance, if malicious or accidental alterations to a traffic sign cause a self-driving car to suddenly accelerate or brake.

In a 2020 paper entitled "A Target-Agnostic Attack on Deep Models: Exploiting Security Vulnerabilities of Transfer Learning," Shahbaz Rezaei and Xin Liu[6] explore how adversarial learning can be applied to transfer learning. Because many modern applications of deep learning use transfer learning, a malicious hacker could analyze the architecture and weights of the pretrained model to exploit weaknesses in the deployed model.

Rezaei and Liu show that a hacker, with knowledge only of the pretrained model used in the application model, can develop adversarial inputs leading to any desired outcome. Because the hacker does not need to know the weights or architecture of the application model or the data it was retrained on, the method of attack is *target agonistic*, which allows it to exploit wide swaths of deployed deep learning models that utilize publicly available pretrained models.

Consider the following model structure, which uses a simple form of transfer learning (Figure 2-25). The input to the model is passed through the pretrained model (both convolutional and FC components). The outputs of the pretrained model (a three-dimensional vector) are passed through a custom (newly added) FC layer, which yields the output (a two-dimensional vector). The weights of the FC layer are labeled – for instance, the connection between the top node of the pretrained model's output and the bottom node of the custom FC layer is 4. For the purposes of simplicity, ignore the bias term in each node.

---

[6] Shahbaz Rezaei and Xin Liu, "A Target-Agnostic Attack on Deep Models: Exploiting Security Vulnerabilities of Transfer Learning," 2019. Paper link: https://arxiv.org/pdf/1904.04334.pdf. Code link (Keras/TensorFlow implementation): https://github.com/shrezaei/Target-Agnostic-Attack.

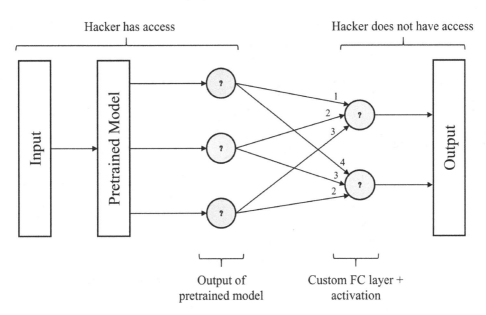

Figure 2-25. *Example transfer learning model architecture with hacker visibility marked*

The hacker has access only to the pretrained model, not the custom FC layer and any operations that process the output of the pretrained model.

Say that the hacker wanted to construct an input that would trigger a certain output neuron, like the bottom neuron in the custom FC layer. The hacker can construct an input such that the top neuron output of the pretrained model is some very large number and all the other output neurons are 0. When multiplied by the corresponding weights, the outputs of the complete model yield 10 and 40; passing through the softmax activation function magnifies this difference, and the bottom output neuron of the complete model is the final decision of the model, as desired, as it has the largest associated value/probability (Figure 2-26).

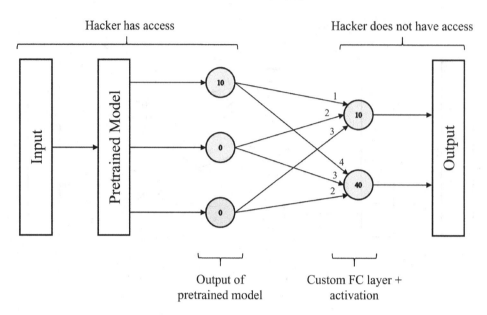

Figure 2-26. *Example transfer learning model architecture with a pretrained model output neuron activated to trigger a specific output from the model*

Alternatively, if the hacker wanted to activate the top output neuron of the complete model, it could design an input such that the output neurons of the pretrained model were all zero except for the bottom one, which would be an arbitrarily large number (Figure 2-27).

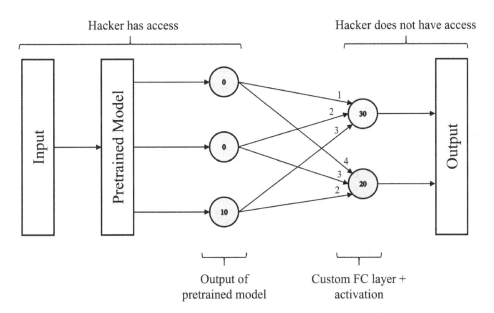

Figure 2-27. *Example transfer learning model architecture with a pretrained model output neuron activated to trigger a specific output from the model*

However, because the hacker does not know the weights outside of the pretrained model in the custom FC layer, they can perform a brute-force search to attempt to force the model to output a certain result. In this search, the hacker aims to construct an input such that the output of the pretrained model is $[a, 0, ..., 0, 0]$ the first time, $[0, a, ..., 0, 0]$ the second time, and so on until the last time, $[0, 0, ..., 0, a]$, where a is some arbitrarily large number. For each desired output, some node in the complete network's output will be activated. The hacker simply records which one of the complete network's output nodes is activated for each designed output of the pretrained model.

It should be noted that for this strategy to be capable of activating every output node of the complete model, for each output node of the complete model, there must exist a connection from an output node of the pretrained model whose weight is larger than the connection from the output node of that pretrained model to any other output node in the complete model. This may not be true in more complex datasets, especially heavy multiclass problems, but in most networks, this strategy works to "force" the network to predict almost any desired result.

Because the problem has now been boiled down to finding the input to the model that will lead to a certain output of the pretrained component, the hacker can use methods like gradient descent to solve the following optimization problem: *adjust the values of the input image such that the output of the pretrained model can as closely match the desired output (minimize error between the output of the pretrained model and the desired output).* In this case, the desired output is some case where all the output nodes except for one are zero, and the one nonzero node is some very large number. The optimization algorithm begins with a standard input image and makes small changes to the image such that the output of the pretrained model is the desired output, constructing an adversarial image. Passing this adversarial image through a model that uses that pretrained model via transfer learning fools it, even though to the human eye no significant changes were made (Figure 2-28).

Figure 2-28. *Left: example input image to the optimization algorithm. Right: example adversarial image generated to produce a certain network output. (By Shahbaz Rezaei and Xin Liu.)*

Making small, specialized distortions to the original input image results in the model making different classifications between the first and second images. Moreover, this could be done without any knowledge of the complete networks' weights or architectures – an impressive feat.

As the number of target classes (of the complete neural network) increases, the performance of the attack method decreases – but it is still alarmingly high. Consider the attack's performance on a neural network using the VGG face pretrained model on the University of Massachusetts Labeled Faces in the Wild (LFW) dataset (Table 2-1).

Table 2-1. *Metrics of attack performance for different numbers of target classes. Target classes: number of target classes/output nodes in the complete neural network. NABAC: number of attempts to break all classes, referring to the number of queries the attack needs to make to "break" a class (force the model to misclassify an item from that class as another class). Effectiveness (x%): the ratio of adversarial inputs that trigger a target class with x% confidence to the total number of adversarial inputs. (By Shahbaz Rezaei and Xin Liu.)*

Target Classes	NABAC	Effectiveness (95%)	Effectiveness (99%)
5	48.25 ± 42.5	91.68% ± 5.69	87.82% ± 6.98
10	149.97 ± 132.15	88.87% ± 2.46	83.07% ± 3.31
15	323.36 ± 253.56	87.79%± 2.42	82.08% ± 2.74
20	413	87.17%	79.16%

As the number of target classes increases, the attack needs to make more queries to break all classes, and its effectiveness decreases.

In the simplified example, only one fully connected layer was added after the pretrained model. While this is one method of integrating the pretrained model into the model structure, often many more layers are added. As the number of layers added afterward increases, performance also decreases (Table 2-2).

Table 2-2. *Metrics of attack performance for different numbers of layers. (By Shahbaz Rezaei and Xin Liu.)*

# New Layers	NABAC	Effectiveness (95%)	Effectiveness (99%)
1	48.25 ± 42.5	91.68% ± 5.69	87.82% ± 6.98
2	51.87 ± 39.94	91.57% ± 4.87	86.45% ± 5.35
3	257.26 ± 387.16	89% ± 8.20	85.67% ± 8.88

Nevertheless, the effectiveness is high, even as the number of target classes and new layers increases. This study is an interesting investigation into the implications of transfer learning for the deployment of models in real-world applications. It's important to keep an eye toward deployment and the model's role in the real world rather than in an airtight experimentation laboratory environment when designing deep learning approaches.

This overview was a simplification of the study; to read more, you can find the paper at `https://openreview.net/pdf?id=BylVcTNtDS`.

Self-Supervised Learning Case Study: Predicting Rotations

In "Unsupervised Representation Learning by Predicting Image Rotations," Spyros Gidaris, Praveer Singh, and Nikos Komodakis[7] propose a simple self-supervised learning method that develops strong representations of the data and yields effective results.

Gidaris et al. propose a pretraining strategy in which images are randomly rotated by 0, 90, 180, or 270 degrees and the model predicts the angle by which the image was rotated. The core intuition behind the design of the rotation-based self-supervised learning pretraining strategy is that the model must first understand the concept of the model depicted in the images in order to identify the rotation that was applied to the image. A successful model, moreover, has a sense of the "natural orientation" of an object, and thus have gravity and other forces should act on objects in an image.

> *…it is essentially impossible for a ConvNet model to effectively perform the above rotation recognition task unless it has first learnt to recognize and detect classes of objects as well as their semantic parts in images. More specifically, to successfully predict the rotation of an image the ConvNet model must necessarily learn to localize salient objects in the image, recognize their orientation and object type, and then relate the object orientation with the dominant orientation that each type of object tends to be depicted within the available images.*

> —Gidaris et al., "Unsupervised Representation Learning by Predicting Image Rotations"

---

[7] Sypros Gidaris, Praveer Singh, and Nikos Komodakis, "Unsupervised Representation Learning by Predicting Image Rotations," 2018. Paper link: `https://arxiv.org/pdf/1803.07728.pdf`. Code link: `https://github.com/gidariss/FeatureLearningRotNet` (PyTorch implementation, but main ideas can be translated by user into Keras/TensorFLow syntax).

With no access to the original, unrotated image, given an image rotated θ degrees (chosen from θ = 0°, 90°, 180°, 270°), the network must maximize the probability that the image is rotated θ degrees (Figure 2-29).

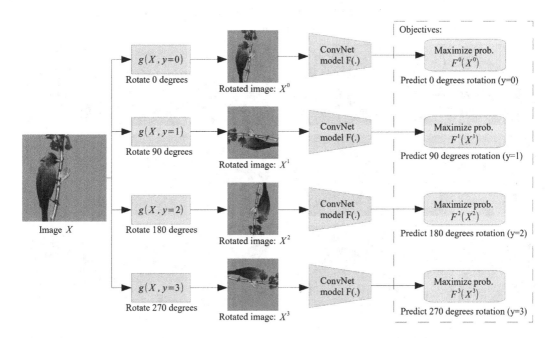

Figure 2-29. *The convolutional neural network's prediction objectives under self-supervised learning by predicting rotations. (By Gidaris et al.)*

To visually see the difference in the learned representations for self-supervised learning as opposed to standard supervised learning (only training the model on the task dataset without constructing a pretraining task), Gidaris et al. visualize the attention maps from a model trained with a standard supervised procedure and another trained with a self-supervised procedure (Figure 2-30).

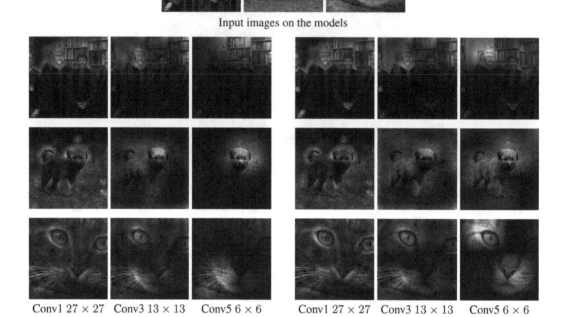

Input images on the models

Conv1 27 × 27   Conv3 13 × 13   Conv5 6 × 6   Conv1 27 × 27   Conv3 13 × 13   Conv5 6 × 6

(a) **Attention maps of supervised model**   (b) **Attention maps of our self-supervised model**

Figure 2-30. *Attention maps from various layers from a supervised model and a self-supervised model given three example input images. (By Gidaris et al.)*

The attention map of a convolutional layer can be found by computing the feature maps for that layer and aggregating them. You can visually see that the attention maps from the self-supervised model focus more on the high-level objects, like the eyes, nose, tails, and heads – in accordance with important features required to perform object recognition.

Note   You can perform something similar by extracting features from a layer in the middle of a neural network in Keras by referring to individual layers (assuming each layer is associated with a variable you defined using the Functional API). Visualizing layer weights and features can be helpful to understanding the improvement in representation capabilities that self-supervised learning provides.

Interestingly, Gidaris et al. find that classifying 0°-90°-180°-270° rotations performs better than other rotations, like 0°-180° rotations, 90°-270° rotations, and 0°-45°-90°-135°-180°-225°-270°-315° (Table 2-3). This result highlights the importance of experimenting with variations of self-supervised strategies to explore which ones perform the best for your particular problem and approach.

Table 2-3. Performance on CIFAR-10 classification for different rotations needed for prediction. (By Gidaris et al.)

# Rotations	Rotations	CIFAR-10 Classification Accuracy
8	0°-45°-90°-135°-180°-225°-270°-315°	88.51
4	0°-90°-180°-270°	**89.06**
2	0°-180°	87.46
2	90°-270°	85.52

We can implement self-supervised learning via predicting rotations relatively easily. We will use the CIFAR-10 dataset, which can be loaded directly from Keras datasets with (x_train, y_train), (x_test, y_test) = keras.datasets.cifar10.load_data(). The first step after loading data is to generate self-supervised altered task data, in this case predicting which degree the image has been rotated (Listing 2-20). Although there are more efficient methods, we will simply loop through every data instance in x_train, randomly rotate it by 0, 90, 180, or 270 degrees, and save the rotated image and corresponding label into a new list. The scipy library's scipy.ndimage.rotate(img, deg) allows us to easily perform rotations.

Listing 2-20. Constructing rotation-based altered task dataset

```
from scipy.ndimage import rotate
ss_x_train, ss_y_train = [], []
for ind in range(len(x_train)):
    rot = np.random.choice([0, 90, 180, 270])
    if rot==0:
        ss_x_train.append(rotate(x_train[ind], rot))
        ss_y_train.append([1,0,0,0])
    elif rot==90:
```

```
        ss_x_train.append(rotate(x_train[ind], rot))
        ss_y_train.append([0,1,0,0])
    elif rot==180:
        ss_x_train.append(rotate(x_train[ind], rot))
        ss_y_train.append([0,0,1,0])
    elif rot==270:
        ss_x_train.append(rotate(x_train[ind], rot))
        ss_y_train.append([0,0,0,1])
ss_x_train = np.array(ss_x_train)
ss_y_train = np.array(ss_y_train)
```

We'll use the EfficientNetB4 architecture (not using its weights) to predict which one of four rotations was applied (Listing 2-21). Note that the output of the EfficientNetB4 architecture is 100-dimensional, which is projected via the final layer out into the four desired outputs. A more efficient approach, at first glance, may seem to simply define the number of output classes in the base EfficientNetB4 model to be 4. We will see soon why this design instead is required for this sort of self-supervised learning operation.

Additionally, we use input_tensor=inp rather than the input_shape=(a,b,c) and base_model(params)(inp) syntax discussed prior. If we were to use the latter method, the variable base would store a tensor (i.e., the output of the model) rather than the model itself. We need to be able to access the EfficientNet base model to transfer it into a new architecture after the self-supervised process has been completed.

Listing 2-21. Constructing model architecture

```
inp = L.Input((32,32,3))
base = EfficientNetB4(input_tensor=inp,
                      weights=None,
                      classes=100)
out = L.Dense(4, activation='softmax')(base.output)
model = Model(inputs=inp,
              outputs=out)
```

The model can be compiled and fitted on the altered task dataset (Listing 2-22). In this case, because the dataset is so small (largely because of the small image resolution rather than the number of items in the dataset), passing numpy arrays suffices and no TensorFlow dataset conversions are needed.

Listing 2-22. Constructing model architecture

```
model.compile(optimizer='adam',
              loss='categorical_crossentropy',
              metrics=['accuracy'])
model.fit(ss_x_train, ss_y_train,
          epochs=100)
```

After fitting on the altered task dataset, though, we run into a problem: the current model outputs only four values, since the altered task dataset had four unique labels, but the task dataset (i.e., the formal CIFAR-10) dataset has ten classes. This requires an architectural modification.

Herein lies the reason behind defining a specialized output layer for the altered task dataset and not specifying `classes=4` in the base EfficientNetB4 model when training on the altered task dataset: we can extract the base model and attach it to a new output layer with ten classes, as follows (Listing 23). Moreover, because we defined `base` to be a model rather than an output tensor, we can connect it via the input easily via `base(inp)`.

Listing 2-23. Constructing model architecture

```
inp = L.Input((32,32,3))
base_out = base(inp)
out = L.Dense(10, activation='softmax')(base_out)
model = Model(inputs=inp,
              outputs=out)
```

If we instead had trained a base model with output four classes and attached another output layer with ten classes afterward, which is another technically available option (in that the code runs), performance would likely be poor due to the bottleneck imposed by the four-class output. In this case, think of the EfficientNet model not as outputting 100 classes but as outputting 100 features that are compiled and used to make decisions in the output layer. The base model still contains all the weights learned from self-supervised learning. The new model can then be compiled and trained on the task dataset.

This self-supervised learning approach drastically improves the state-of-the-art results on unsupervised feature learning for a variety of object detection tasks, including ImageNet, PASCAL, and CIFAR-10.

The work of Gidaris et al. points to the existence of simple yet effective approaches to apply self-supervised learning to boost deep learning model performance.

Self-Supervised Learning Case Study: Learning Image Context and Designing Nontrivial Pretraining Tasks

Earlier, we discussed relatively simpler self-supervised learning examples, like adding noise to the task dataset and pretraining the model to predict the degree of noise. Other examples were simple conceptually (although perhaps difficult to implement or accomplish), like pretraining the model to colorize grayscale images. While sometimes simpler self-supervised strategies like predicting the degree of rotation are effective, in other occasions for a bigger performance boost, a more complex pretraining task needs to be used.

Carl Doersch, Abhinav Gupta, and Alexei A. Efros propose a complex, creative, and effective self-supervised learning pretraining task in their 2016 paper, "Unsupervised Visual Representation Learning by Context Prediction."[8] Their design is a great case study of the implications and possibilities that need to be considered to design a successful self-supervised pretraining strategy.

In the context prediction pretraining task, the model must predict the relative location, or the context, of one patch to another (Figure 2-31). Patches are small, square-shaped regions of the image. The model is fed a base patch (outlined in blue) and a neighboring patch selected from the eight regions around the base patch (outlined in dotted red) and trained to predict the location of the neighboring patch in relation to the base patch. In the example, the label would be "top left."

---

[8] Carl Doersch, Abhinav Gupta, and Alexei A. Efros, "Unsupervised Visual Representation Learning by Content Prediction," 2016. Paper link: https://arxiv.org/pdf/1505.05192.pdf. Code link: https://github.com/cdoersch/deepcontext (implementation in Caffe, but ideas can be translated into Keras/TensorFlow syntax).

Example:

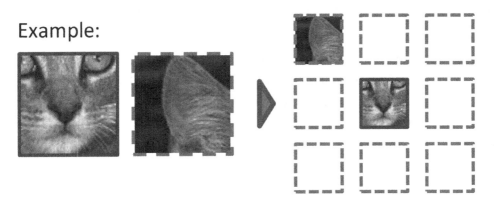

Figure 2-31. *Example context and patch relationships*

Each general direction that could describe the location of a neighboring patch relative to the base patch is associated with a number. Thus, the model takes in two images (the base patch and the neighboring patch) and outputs a vector indicating which region relative to the base patch the neighboring patch is most likely to be located (Figure 2-32).

Figure 2-32. *Assigning numbers to neighbor patch directions*

To perform this pretraining task, Doersch et al. constructed a multi-input network that takes in two inputs, processes the images through a series of image-based layers in separate "branches," merges the two branches, processes the merged results, and outputs the location of one input relative to the other (Figure 2-33). Dotted lines in Figure 2-33 indicate shared weights, which means that the weights for the two layers are the same. We'll discuss multi-input networks, branching, and weight sharing in later chapters.

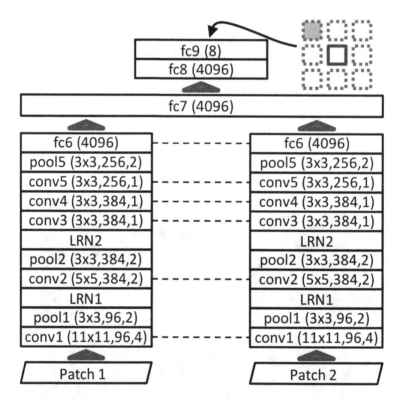

Figure 2-33. *Two-input architecture used to perform the pretraining task*

This pretraining task, at least in theory, requires a deep understanding of the core features of certain items and the high-level relationships between attributes of objects. This makes it difficult, even for humans. Try answering these two visual questions (Figure 2-34).

Question 1:   Question 2:

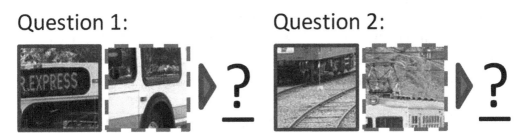

Figure 2-34. *Example "questions" with two patch pairs of inputs*

However, Doersch et al. found that the neural network was learning "trivial solutions" – that is, it was finding shortcuts and easy approaches to the pretraining task that were not related to the concepts Doersch et al. had hoped it would learn.

For instance, low-level cues like boundary patterns or the continuation of textures across patches serve as trivial shortcuts that the neural network can exploit without learning the high-level relationships it would ideally learn. To address this, Doersch et al. added both a gap between patches about half the length of the patch width and a random "jitter" for each patch location by up to 7 pixels. (These are shown visually in Figure 2-32.) With these modifications, the network could not reliably exploit trivial shortcuts like the continuation of textures and lines.

However, another unexpected problem arose: a phenomenon called chromatic aberration, in which light rays pass through a lens focus at different points depending on their wavelength. Thus, in camera lenses that are affected by chromatic aberration, one color – usually green – is "shrunk" toward the center of the image relative to other color channels (red and blue). The network can thus parse the absolute location of patches by detecting the shape and position of the separation between green and magenta (the color comprising the non-green color channels, red and blue). Finding the relative position given knowledge about the absolute location of two patches thus becomes a trivial task.

To address this, Doersch et al. randomly dropped two of three color channels from each patch and replaced the dropped channels with Gaussian noise. This method prevents the neural network from exploiting information in the color channels and instead to focus on the general shapes and edges – the higher-level features – in its pretraining task.

By making modifications to the architecture, like only using one "branch" and adding more convolutional layers, Doersch et al. converted the two-image-to-vector architecture into an image-to-vector architecture. The authors tested this pretrained architecture on the PASCAL Visual Object Classes (VOC) Challenge and found that this self-supervised learning strategy was the best result without using labels not included in the dataset (to their knowledge, and at the time of paper release). The authors found in other datasets that this context prediction task yielded a significant boost to model performance.

This paper is both an example of a well-thought-out self-supervised learning pretraining task and of the process of modifying the pretraining task to accomplish the ultimate objective by discouraging trivial solutions. Self-supervised learning offers us a clever way to find, extract, and utilize rich insights and information without needing to search for costly labels and wandering outside of the dataset.

Key Points

In this chapter, we discussed the intuition and implementation behind pretraining, transfer learning, and self-supervised learning:

- Pretraining is a task that is performed before the formal training of a model and can orient the model toward better performing on its ultimate task.

- Because pretraining moves the model into a more convenient location in the loss landscape, it can arrive at a better solution more quickly than a model without pretraining. Thus, pretraining generally has two key advantages that boost model performance: decreased time and better metric scores.

- Transfer learning uses two separate datasets – a general dataset and a task dataset. The model is first trained on the general dataset, and then the learned weights and "skills" from the task are transferred to another model that is trained on the task dataset. Pretrained models are often kept in publicly available repositories and can be accessed and used in your own model.

- Image-based pretrained models generally consist of two components: a feature extraction component and a fully connected component (the "top"). You can (a) keep only the feature extraction component and build a custom FC component, (b) keep both the feature extraction component and the fully connected component only, or (c) keep both the feature extraction component and the FC component, but add more layers afterward.

- Layer freezing allows you to selectively train certain parts of a model using transfer learning to better take advantage of pretrained weights. Usually, weights from the pretrained model are frozen and the custom added layers are trained. After this step, sometimes the pretrained model is unfrozen and the entire network is trained to fine-tune on the data.

- Keras/TensorFlow has built in a repository of pretrained models organized into modules, in which each module contains the pretrained model(s) and associated functions (usually for encoding and decoding). Important pretrained models include the ResNet family, the InceptionV3 model, the MobileNet family, and the EfficientNet family. Keras/TensorFlow models are pretrained on the ImageNet dataset. You can also convert PyTorch models into Keras/TensorFlow models to take advantage of PyTorch's large repository of pretrained models.

- Self-supervised learning uses pretrains on an altered task dataset derived from input data of the task dataset. The x and y of the altered task dataset come completely from only the x of the task dataset, often by performing distortions or actions to generate labels. Self-supervised learning helps orient the model to representing and understanding the "world" of the dataset so it is prepared to understand the dynamics and core features of the task dataset.

  - Self-supervised learning is especially valuable for training on small datasets with few labels, or when there is an abundance of unlabeled data that otherwise would not have been used.

- There are two key differences between transfer learning and self-supervised learning. Firstly, the pretraining dataset and the task dataset generally come from completely different sources in transfer learning, whereas in self-supervised learning, the pretraining dataset is derived from the input data of the task dataset. Secondly, transfer learning aims to develop "skills," while self-supervised learning encourages "representations of the world."

Although we implemented relatively simple models and methods in this chapter, with these conceptual tools and the design intuition of what transfer learning and self-supervised learning are, you will be able to build many more complex training structures, like stacking pretrained models ("double transfer learning") or designing innovative self-supervised pretraining methods.

In the next chapter, we'll expand upon our study of advanced neural network structures and manipulations with autoencoders, a versatile tool often used in developing successful training structures.

CHAPTER 3

The Versatility of Autoencoders

One word is worth a thousand pictures, if it's the right word.

—Edward Abbey, American author and essayist[1]

The concept of encoding and decoding entities – ideas, images, physical material, information, and so on – is a particularly profound and important one, because it is so deeply embedded into how we experience and understand the environment around us. Encoding and decoding information is a key process in communication and learning. Every time you communicate with someone, observe the weather, read from a book – like this one – or in some way interact with information, you are engaging in a process of encoding and decoding, observing and interpreting. We can make use of this idea with the deep learning concept of an autoencoder.

This chapter will begin by exploring the intuition and theory behind autoencoders and how they can be used. It will also cover Keras neural network manipulation methods and neural network design that are integral to implementing not only autoencoders but also other more complex neural network designs in later chapters. The other major half of the chapter will be dedicated toward exploring the versatility of autoencoders through five applications. While the autoencoder is a simple concept, its incorporation into efforts tackling deep learning problems can make the difference between a mediocre and a successful solution.

---

[1] Edward Abbey (1984). "Beyond the Wall: Essays from the Outside," p. 11, Macmillan.

Autoencoder Intuition and Theory

Although we will clarify what encoding and decoding refer to more specifically with relevance to the intuition behind autoencoders, for now they can be thought of as abstract operations tied inextricably together to one another. To encode an object, for our purposes, is to represent it at a loss of quantifiable information. For instance, we could encode a book by summarizing it, or encode an experience with key sensory aspects, like notable senses of touch or hearing. To decode is to take in the encoded representation of the object and to reconstruct the object. It is the objective of encoding and decoding to "work together" such that the reconstruction is as accurate as possible. An effective summary (encoding) of a book, for instance, is one from which a reader could more or less construct the main ideas of the book to a high accuracy.

Suppose someone shows you an image of a dog. They allow you to look at that image for a few seconds, take it away, and ask you to reconstruct the image by drawing it. Within those few seconds, ideally you would have extracted the key features of that image so that you could draw it as accurately as possible through some efficient method. Perhaps you remember that the dog was facing right, its head upright, that it was tall, and that it looked like it was standing still.

What is interesting is that these are all high-level abstract concepts – to deduce that an object is facing a certain direction, for instance, you must know the relationship between the directions of each of the object's parts (for instance, if the head is facing one way, the tail should be facing the other) and be able to represent them in a spatial environment. Alternatively, knowing that a dog is tall requires knowledge of what a "standard" or "short" dog looks like. Identifying the absence of motion requires knowledge of what motion for this particular class of objects looks like.

The key idea here is that for effective compression and reconstruction of an object in the context of complex objects and relationships, the object must be compressed with respect to other objects. Compressing an object in isolation with a high reconstruction performance is more difficult or performs worse than compressing an object with access to additional sets of related knowledge. This is why recovering an image of a dog is easier than recovering an image of an unfamiliar object. With the image of the dog, you first identify that the object is a dog and thus identify that knowledge relating to the "dog" entity is relevant and then can observe how the image deviates from the template of the dog. When decoding, if you already know that the image was of a dog (a technical detail: this information can be passed through the encoding), you can do most of the heavy lifting by first initializing the knowledge related to a dog – its standard anatomy, its behavior, and its character. Then, you can add deviations and specifics, as necessary.

Because efficient compression of complex entities requires the construction of highly effective representations of knowledge and of quick retrieval, this encoding-decoding pair of operations is extremely useful in deep learning. This pair of operations is referred to generally as the autoencoder structure, although we will explore more solidly what it entails. Because the processes of encoding and decoding are so fundamental to learning and conducive to the development of effective representations of knowledge, autoencoders have found many applications in deep learning.

Likewise, it should be said that this sort of dependence on context makes autoencoders often a bad approach for tasks like image compression that require a more universal method of information extraction. Autoencoders are limited as tools of data compression to the context of the data they are working with and even with that context perform at best equivalently to existing, more universal compression algorithms. For instance, you could only train a standard autoencoder to effectively encode a few pineapple images if you fed it thousands of other images of pineapples; the autoencoder would fail to reconstruct an image of an X-ray, for instance. This is a good example of critically evaluating deep learning's viability as an approach to certain problems and a much needed reminder that deep learning is not a universal solution and needs to work with other components to produce a complete product.

Consider the following reconstructions of images in the MNIST digit dataset by an autoencoder that was forced to encode each image into only four numbers and then to decode the image from those four numbers (Figure 3-1). We can observe that the network relies upon "templates" of digits. For instance, the original digit "5" has a very distinct shape, with sharp edges and joints, but the network's reconstruction looks like a different image – although it expresses the same concept of "5," it is much more rounded and less stylistic. Similar patterns can be observed for the digits "0," "4," "1," and "9." Here, we see that the autoencoder is not merely finding an efficient compression algorithm for the images but is performing clustering and a bit of under-the-hood "soft" classification to help reconstruct the general concepts of the input image.

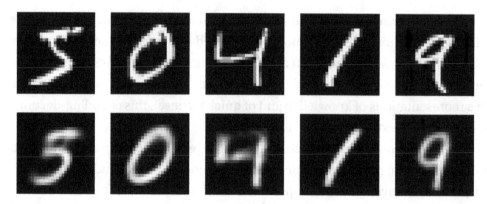

Figure 3-1. *Top row: inputs/original images. Bottom row: reconstructed inputs*

Despite the autoencoder being an unsupervised task – there are no labels – it is able to perform highly effective clustering and learn key features, attributes, and structures of the data.

Formally, the autoencoder consists of three general components: the encoder, the bottleneck, and the decoder (Figure 3-2). These components are not exclusive from one another; they overlap. The encoder takes in the input and outputs the encoded representation, and the decoder takes in the encoded representation and decodes it into the output. The "bottleneck" refers to the nodes shared between the encoder and the decoder: it is the output of the encoder and the input to the decoder. It can be thought of as holding the encoded representations of the input. It is also known as the latent space, "latent" coming from the Latin word for "to be hidden" – it is a space containing "hidden" representations for compressed/encoded data decipherable only by the decoder.

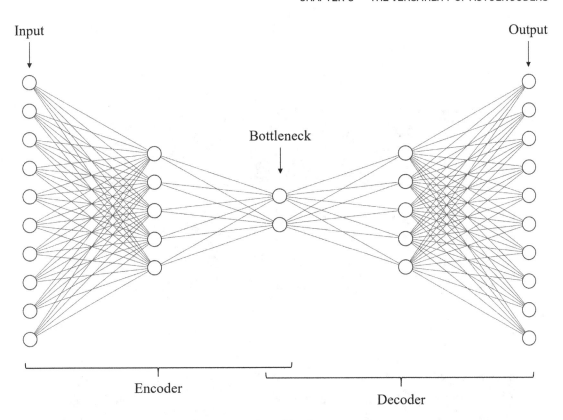

Figure 3-2. *Components of an autoencoder diagram*

The input and output of autoencoders have the same size, because autoencoders are trained on the task of reconstruction – given an input item, the objective is to reconstruct the output with as high accuracy as possible. This is a crucial feature of autoencoders that will be core to how we go about designing these structures.

Training a neural network to "do nothing" – to output whatever the input is – is a trivial task for many architectures. What makes the autoencoder so meaningful and useful for a wide variety of applications, though, is the bottleneck. The decrease in the number of nodes in the bottleneck layer compared to the number of nodes needed to represent the input data in its original form (or substitute nodes with any other unit of data storage) makes the task of reconstruction difficult and forces the autoencoder to develop efficient encoding and decoding methods. Figures 3-3 and 3-4 show the reconstructed representations for two autoencoders trained on the MNIST digit dataset, one with a larger bottleneck than the other. While the first autoencoder clearly had too small of an autoencoder to develop reasonably accurate reconstructions, the other autoencoder is able to develop a certain level of detail – the pointiness of the five, the slight curl of the nine's stem – that it may warrant concern as to if the task is too trivial. This depends on the application you are using the autoencoder for, though.

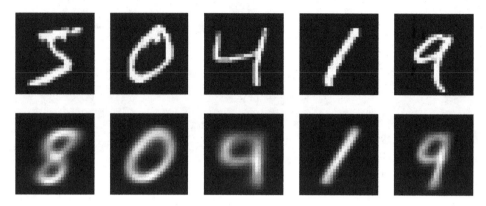

Figure 3-3. *Original images and their reconstructions by an autoencoder with a small bottleneck size*

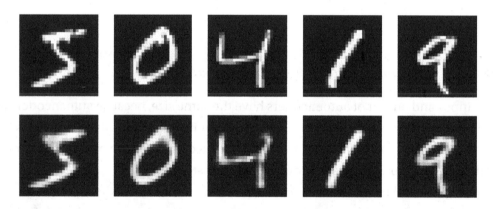

Figure 3-4. *Original images and their reconstructions by an autoencoder with a large bottleneck size*

Most of the "meat" within autoencoder usage resides in their applications. We'll begin by discussing how autoencoders are designed and implemented for tabular data and image data. Then, we'll explore the intuition and implementation for five applications of autoencoders.

The Design of Autoencoder Implementation

There are many considerations to be made in the implementation of autoencoders, depending on the form of their application. In this section, we will broadly discuss considerations for designing autoencoders on one-dimensional (vector) and two-dimensional (tabular) data by example.

Autoencoders for Tabular Data

Tabular data, for our purposes, refers to data that can be put in the form (n, s), where n indicates the number of data points and s indicates the number of elements in each data point. The entire dataset can be put into a two-dimensional table, where each row indicates a specific data point and each column indicates a feature of that data point.

Building autoencoders for tabular data is an appropriate place to begin, because the shape of data at any one layer can be easily manipulated simply by changing the number of nodes at that layer. The shape of other forms of data, like images, is more difficult to handle. Being able to manipulate shape is crucial to autoencoder design because the concept of encoding and decoding entities comes with relatively strict instructions about what the shape of the data should look like before encoding, after encoding, and before decoding.

The difference between a standard autoencoder (often abbreviated simply as "AE") and a "deep autoencoder" should also be noted here. Although the exact terminology has not seemed to have stabilized yet, generally an autoencoder refers to a shallow autoencoder structure, whereas a deep autoencoder contains more depth. Often, this is determined by the complexity of the input data, since more complex forms of data usually necessitate greater depth. Because data that can be arranged in a tabular format is generally less complex than image or text data (or other sorts of highly specialized and complex data that neural networks have become good at modeling), it's generally safe to refer to autoencoder structures used on high-resolution sets of images or text as deep autoencoders and autoencoder structures used on tabular data as simply autoencoders.

In this example, we will be building an autoencoder with 784-dimensional data. That is, each data point consists of 784 features. Let's begin by importing necessary models and modules. We'll use the `import keras.layers as L` method of dealing with importing layers instead of importing each layer individually by name out of convenience (Listing 3-1). Additionally, because there is no need to use the Functional

API in this case – we are not building a nonlinear topology and do not need such strong access to some layer as to warrant assigning it its own variable – we will opt for the simpler Sequential model structure.

Listing 3-1. Importing important layers and models

```
import keras.layers as L
from keras.models import Sequential
```

Once we initialize the Sequential model structure, we can add the Input layer, which takes in the 784-dimensional input data.

Now, we will build the encoding component of the autoencoder (Listing 3-2). The encoder should successively decrease the size the data takes in as to guide the process of encoding. In this case, the first decrease in size is from 784 nodes to 256 nodes; the second is from 256 nodes to 64 nodes; the third is from 64 nodes to 32 nodes. By decreasing the number of nodes in each Dense layer, we are decreasing how much space we give the autoencoder to represent the input data.

Listing 3-2. Building a simple encoder using Dense layers for tabular data

```
model = Sequential()
model.add(L.Input((784,)))
model.add(L.Dense(256, activation='relu'))
model.add(L.Dense(64, activation='relu'))
model.add(L.Dense(32, activation='relu'))
```

The last layer of the encoder contains 32 nodes, indicating that the bottleneck will be 32 nodes wide. This means that, at its most extreme, the autoencoder must find a representation of 784 features with 32 values.

It should also be noted that it is convention for the number of nodes in each layer to be a power of two. You will see this pattern both throughout examples in this book and in the architectures of neural networks designed by researchers (some of these are presented in case studies). It's thought by many to be convenient for memory and a good way to scale the number of nodes meaningfully (when the number of nodes is high, meaningful change is *proportional* rather than additive). This convention by no means is required, though, if your design requires node quantities that cannot accommodate this convention.

We can then add the decoder layers, which should successively expand the space the neural network has to decode data from its encoded representation (Listing 3-3). Although it's not strictly necessary for the success of an autoencoder, often decoders are built symmetrically to the encoder out of convenience. Here, the decoder performs the same steps as the encoder, but "in reverse." The first expansion is from the encoded representation of 32 nodes into 64 nodes; the second is from 64 nodes to 256 nodes; the last is from 256 nodes to 600 nodes, the shape of the original input data.

Listing 3-3. Building a simple decoder using Dense layers for tabular data

```
model.add(L.Dense(64, activation='relu'))
model.add(L.Dense(256, activation='relu'))
model.add(L.Dense(784, activation='sigmoid'))
```

Note that, in this example, the activation for the last layer of the model is the sigmoid activation function. We traditionally put the sigmoid activation (or other related curved and bounded functions and adaptations) on the last layer to bound the neural network for classification problems. This may or may not be suitable toward your particular task.

If your input data consists of entirely binary data or can be put into that form appropriately, sigmoid may be an appropriate activation. It is important to make sure that your input data is scaled properly; for instance, if a feature is binary in that it contains only the values 10 and 20, you would need to adjust the feature such that it consists only of the values 0 and 1. Alternatively, if your feature is not strictly binary but tends to cluster around two bounds, sigmoid may also be an appropriate choice.

On the other hand, if a feature is spread relatively uniformly across a wide range, it may not be appropriate to use the sigmoid activation. The sigmoid function is sloped such that it is more "tricky" to output an intermediate value near 0.5 than near 0 or 1; if this does not adequately represent the distribution of a feature, there are other options available. An activation like ReLU (Rectified Linear Unit, defined as $y = \max(x, 0)$) may be more appropriate. However, if your feature ranges across negative outputs as well, use the linear activation (simply $y = x$). Note that depending on the character of the features in tabular data, you will need to choose different losses and metrics – the primary consideration being regression or classification.

A challenge of autoencoders for tabular data is that often tabular data is not held together by a unifying factor of context in the same way that image or text is. One feature may be derived from a completely different context than another feature, and thus you may simultaneously have a continuous feature and a categorical feature that a tabular autoencoder must both accommodate. Here, feature engineering (e.g., encoding categorical features to continuous values) is needed to "unify" the problem types of the features.

Plotting the model with plot_model() shows us a successive decrease in size during the encoding step and a successive increase in size during the decoding step, as expected (Figure 3-5).

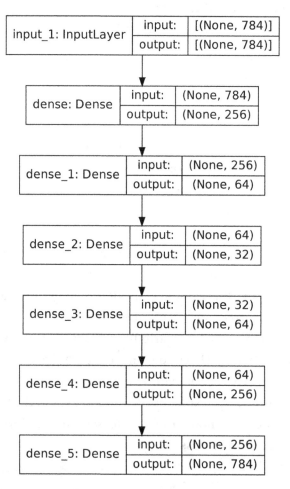

Figure 3-5. *Sample autoencoder using only Dense layers for tabular data*

When building autoencoders, however, generally a *compartmentalized* design is preferred over sequentially stacking layers. A compartmentalized design refers to implementing a model as a relationship between several sub-models. By using a compartmentalized design, we can easily access and manipulate the sub-models. Moreover, it is more clear how the model functions and what its components are. In an autoencoder, the two sub-models are the encoder and the decoder.

To build with a compartmentalized design, first define the architectures of each of the sub-models. Then, use the Functional API to define each sub-model as an input of another object and aggregate the sub-models into an overarching model using `keras.models.Model`.

Let's build the same architecture as we did before, but with a compartmentalized design (Listing 3-4). Note that you can pass the `name='name'` argument into a model to attach a name to it for later reference.

Listing 3-4. Building a simple encoder and decoder using Dense layers for tabular data with compartmentalized design

```
encoder = Sequential(name='encoder')
encoder.add(L.Input((784,)))
encoder.add(L.Dense(256, activation='relu'))
encoder.add(L.Dense(64, activation='relu'))
encoder.add(L.Dense(32, activation='relu'))
decoder = Sequential(name='decoder')
decoder.add(L.Input((32,)))
decoder.add(L.Dense(64, activation='relu'))
decoder.add(L.Dense(256, activation='relu'))
decoder.add(L.Dense(784, activation='sigmoid'))
```

Once we've defined the sub-models (being `encoder` and `decoder`), we can treat them as layers and use the Functional API to aggregate them into an overarching model (Listing 3-5). Recall that we can treat models just like layers in the Functional API by writing them as functions of the previous object.

Listing 3-5. Compiling sub-models into an overarching model – the autoencoder – with compartmentalized design and the Functional API

```
ae_input = L.Input((784,), name='input')
ae_encoder = encoder(ae_input)
ae_decoder = decoder(ae_encoder)

from keras.models import Model
ae = Model(inputs = ae_input,
           outputs = ae_decoder)
```

When plotting the architecture of the overarching model (Figure 3-6), Keras now displays only the direct layers and sub-models that compose it. This can be helpful for understanding the general architecture and data flow of an autoencoder with many layers without needing to see the specifics of what each individual layer does.

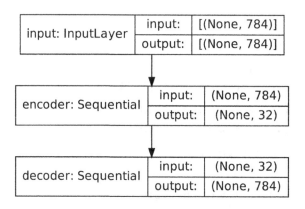

Figure 3-6. *Visualization of an autoencoder architecture built with compartmentalized design*

The primary practical benefit of using compartmentalized design for autoencoders is that after compiling and fitting, we can call encoder.predict(input_data) to obtain the learned encoding. If you do not use compartmentalized design, you can also use layer-retrieving methods discussed in Chapter 2 (e.g., get_layer()) to create a model object consisting of the encoding layers, but it's more work than is necessary and is less portable. Accessing the encoder's encoding of data is necessary for many of the autoencoder applications we will discuss in the second half of this chapter.

As mentioned in Chapter 2, you can use this implementation design method with all sorts of other structures beyond autoencoders to easily freeze entire groups of layers, or for the other benefits of compartmentalized design mentioned here, like better organization or easier referencing of models. For instance, this sort of compartmentalized design can be used to separate a model using transfer learning into two sub-models: the pretrained convolutional component and the custom fully connected layer. Freezing the pretrained convolutional component can be easily done by calling submodel.trainable = False.

Autoencoders for Image Data

Building autoencoders for image data follows the same logic as building one for tabular data: the encoder should condense the input into an encoded representation by using "reducing" operations, and the decoder should expand the encoded representation into the output by using "enlarging" operations. However, we need to make additional considerations to adapt to an increased complexity of the data shape.

The "enlarging" operation needs to be some sort of an "inverse" of the "reducing" operation. This was not much of a concern with Dense layers, because both an enlarging and a reducing operation could be performed simply by increasing or decreasing the number of nodes in a following layer. However, because common image-based layers like the convolutional layer and the pooling layer can only be reductive operations, we need to explicitly note that the decoding component is not only the encoding component "in reverse" (as was described in building autoencoders for tabular data) but is inverting – step by step – the encoding operations. This poses complications to building the encoder and decoder.

Although there are developments on using deep autoencoders for language and advanced tabular data, autoencoders have primarily been used for image data. Because of this, an extensive knowledge of autoencoders is necessary to successfully deal with most image-related deep learning tasks.

Image Data Shape Structure and Transformations

Because shape is so important to convolutional autoencoder design, first, we must briefly discuss image shape and methods of transforming it.

An image has the shape (a, b, c). Using the commonly used channels-last notation, a and b represent the spatial dimensions of the image and c represents the number of channels or the depth (Figure 3-7). For instance, a standard color image has three channels, corresponding to the red, green, and blue values in RGB images. Image-based layers can generally manipulate an image by altering all three elements of its shape.

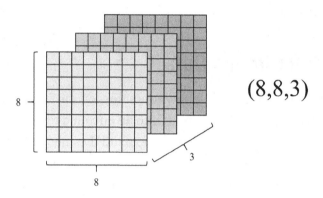

Figure 3-7. *Illustration of the three dimensions of an image*

Table 3-1 enumerates four key layers used in convolutional neural networks. Note that usually the height and the width of the image are the same (the image is square), which makes exploring transformations simpler. This table assumes a square shape, although you can apply similar logic to rectangular inputs. Moreover, the table assumes a stride of 1 is used and that the inputs are spatially two-dimensional, although again the logic applies to other forms.

Table 3-1. *Important layers in a convolutional neural network and their transformations to the image shape*

Layer	Parameters	Output given input shape `(a, a, c)`	Description

Convolution	Kernel shape = `(x, y)`
	 Number of filters = n
	 Padding = "same" or "valid"
		If padding = `'valid'`:
		 `(a-(x-1),`
		 `a-(x-1), n)`
		 If padding = `'same'`:
		 `(a, a, n)`
			The two-dimensional convolution slides a kernel of shape `(x, y)` across the image. Hence, it will reduce the image's spatial dimensions by x-1 and y-1 pixels. Generally x=y (thus the kernel is square-shaped), but some architectures exploit a rectangular kernel shape, which can be successful in certain domains. See the Chapter 6 case study on the InceptionV3 architecture to explore convolution factorization and nonsquare kernel shapes. However, if padding is set as "same," the image is padded (extra "blank" dimensions are added to its size) such that when the convolution is performed, the resulting image has the same shape as the previous one. We'll see why this is helpful later.

(continued)

Table 3-1. (*continued*)

Layer	Parameters	Output given input shape (a, a, c)	Description
Pooling	Pooling size = (x, x) Padding = "same" or "valid"	If a is divisible by x and padding='same' or 'valid': (a/x, a/x, c) If a is not divisible by x and padding='valid': (floor(a/x), floor(a/x), c) If a is not divisible by x and padding='same': (ceil(a/x), ceil(a/x), c)[2]	The two-dimensional pooling operation offers a faster way to reduce the size of an image shape by placing non-overlapping (unlike convolutions, which do overlap) windows of size (x, x) across the image to "summarize" the important findings. Pooling is usually either used in the form of average pooling (all elements in the pooling window are averaged) or max pooling (the maximum element in the pooling window is passed on). Pooling divides the image size by the pooling window's respective dimension. However, because images sometimes may not be exact multiples of the window's respective dimension, you can use padding modes to help determine the exact shape of the output.

(*continued*)

---

[2] "floor" and "ceil" refer to rounding to the lower and upper integer, respectively. For instance, the floor of 10/3 is 3, whereas the ceiling is 4. Therefore, if you were to use pooling with padding = 'same' on an image with shape (10,10,3), the output would be (3,3,3); if you used padding = 'valid', the output would be (4, 4, 3).

Table 3-1. (*continued*)

Layer	Parameters	Output given input shape (a, a, c)	Description
Transpose convolution	Kernel shape = (x, y) Number of filters = n Padding = "same" or "valid"	If padding = 'valid': (a+(x-1), a+(x-1), n) If padding = 'same': (a, a, n)	The transpose convolution can be thought of as the "inverse" of the convolution. If you passed an input through a convolutional layer and then a transpose convolutional layer (with the same kernel shape), you'd end up with the same shape. When you're building the decoder, use transpose convolutional layers in lieu of the convolutional layers in the encoder to increase the size of the image shape. Like the convolutional layer, you can also choose padding modes.

(*continued*)

Table 3-1. (*continued*)

Layer	Parameters	Output given input shape (a, a, c)	Description
Upsampling	Upsampling factor: (x, y)	(a*x, a*y, n)	The upsampling layer simply "magnifies" an image by a certain factor without changing any of the image actual values. For instance, the array [[1, 2], [3, 4]] would get upsampled simply as [[1, 1, 2, 2], [1, 1, 2, 2,], [3, 3, 4, 4], [3, 3, 4, 4]] with an upsampling factor of (2,2). The upsampling layer can be thought of as the inverse of the pooling operation – while pooling divides the dimensions in the image size by a certain quantity (assuming no padding is being used), upsampling multiples the image size by that quantity. You cannot use padding with upsampling. When you're building the decoder, use upsampling layers in lieu of pooling layers in the encoder to increase the size of the image shape.

Note that only convolutional and transpose convolutional layers contain weights; pooling and upsampling are simple ways to aggregate extracted features without any particular learnable parameters. Additionally, note that the default padding method is "valid."

There are many approaches to building convolutional autoencoders. We'll cover many approaches, beginning with a simple convolutional autoencoder without pooling to introduce the concept.

Convolutional Autoencoder Without Pooling

As noted before, autoencoders are generally built symmetrically, but this is even more true with image-based autoencoders. In this context, *not* building the autoencoder symmetrically requires a lot of arduous shape tracking and manipulation.

Let's begin with building an encoder that takes in data with shape (256, 256, 3) and successively encodes it with convolution layers (Listing 3-6). In many convolutional autoencoder designs, the number of filters increases as the image size decreases. Make sure that the increase in number of filters does not outweigh the decrease in image size in terms of the amount of storage the network has available, such that the storage capacity for data decreases throughout the encoder (and correspondingly increases throughout the decoder).

Listing 3-6. Building an encoder for image data using convolutions without pooling

```
encoder = Sequential()
encoder.add(L.Input((64, 64, 3)))
encoder.add(L.Conv2D(8, (3, 3)))
encoder.add(L.Conv2D(8, (3, 3)))
encoder.add(L.Conv2D(16, (3, 3)))
encoder.add(L.Conv2D(16, (3, 3)))
encoder.add(L.Conv2D(32, (3, 3)))
encoder.add(L.Conv2D(32, (3, 3)))
```

In this case, we are successively increasing the number of filters (from 3 channels initially to 8, 16, and 32) while keeping the filter size at (3,3).

By visualizing with plot_model (Figure 3-8), we see how the shape of the image changes over time in accordance with the convolutional layers. The output of the encoder is a compressed image of shape (52, 52, 32). However, you'll notice that this model violates the requirement that the number of filters shouldn't outweigh the decrease in image size without pooling – the "encoded representation" is larger than the input (64*64*3 < 52*52*32)! This is because convolutions are simply not a good way to reduce the image size quickly. We need pooling to address this problem. For now, to keep things simple, we will ignore this issue.

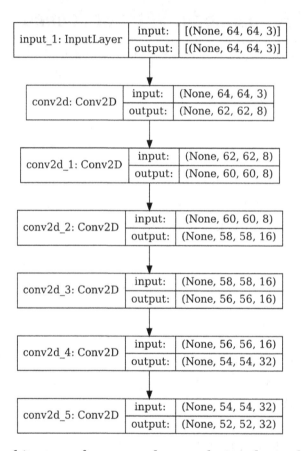

Figure 3-8. *The architecture of an example convolutional encoder without pooling and only using convolutional layers*

We can build our decoder to "mirror" the encoder by reversing the layer order and using the inverse layers to "undo" the encoding (Listing 3-7, Figure 3-9). We know that the decoder will take in an encoded representation of size (52, 52, 32).

Listing 3-7. Building a decoder for image data using convolutions without pooling

```
decoder = Sequential()
decoder.add(L.Input((52, 52, 32)))
decoder.add(L.Conv2DTranspose(32, (3, 3)))
decoder.add(L.Conv2DTranspose(32, (3, 3)))
decoder.add(L.Conv2DTranspose(16, (3, 3)))
decoder.add(L.Conv2DTranspose(16, (3, 3)))
decoder.add(L.Conv2DTranspose(8, (3, 3)))
decoder.add(L.Conv2DTranspose(8, (3, 3)))
```

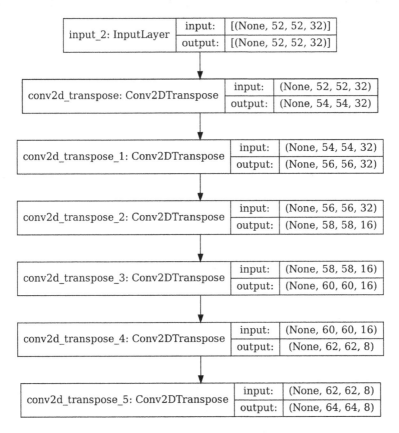

Figure 3-9. *The architecture of an example convolutional decoder without pooling and only using convolutional layers*

There's one problem – the output of the decoder has shape (64, 64, 8), whereas the input has shape (64, 64, 3). There are two ways of addressing this. You could change the last layer to L.Conv2DTranspose(3, (3, 3)) such that it has three channels. Alternatively, you could add another layer to the end of the decoder: L.Conv2DTranspose(3, (1, 1)). Because it has a filter size of (1,1), the image width and height are not changed, but the number of channels is collapsed from 8 into 3.

Because convolutional autoencoders (and networks in general) rely on repeated convolutions and other operations, it's generally good practice to build them using a for loop and index lists of parameters as necessary. For instance, we could rewrite the encoder more efficiently as follows (Listing 3-8).

Listing 3-8. Building long, repeated architectures using for loops and lists of parameters accordingly referenced within the loop

```
num_filters = [8, 16, 32]

encoder = Sequential()
encoder.add(L.Input((64, 64, 3)))
for i in range(3):
    encoder.add(L.Conv2D(num_filters[i], (3,3)))
    encoder.add(L.Conv2D(num_filters[i], (3,3)))
```

In this case, since there is a clear pattern with the number of filters in each layer, you could even write it without indexing a parameter list (Listing 3-9).

Listing 3-9. Building long, repeated architectures using for loops without lists of parameters accordingly referenced within the loop

```
encoder = Sequential()
encoder.add(L.Input((64, 64, 3)))
for i in range(3):
    encoder.add(L.Conv2D(2**(i+3), (3,3)))
    encoder.add(L.Conv2D(2**(i+3), (3,3)))
```

A primary benefit of this sort of design is that you can easily extend the depth of the network simply by increasing the number of iterations layer-adding code is looped through, saving you from needing to type a lot of code manually.

Convolutional Autoencoder Vector Bottleneck Design

Often, the bottleneck (the output to the encoder and input to the decoder) is not left as an image – it's usually flattened into a vector and reshaped into an image. Their primary benefit from this is that we are able to obtain vector representations of images that are independent from any spatial dimensions, which makes them more "clean" and easy to work with. Moreover, they can be more easily used with applications like pretraining (more on this later).

To do this, we need to add a Flatten layer at the end of the encoder (Listing 3-10, Figure 3-10). Since flattening an image usually leads to a very large vector, it's common to apply some Dense layers to further process and reduce its size. Another alternative is the GlobalAveragePooling2D or GlobalMaxPooling2D layer, which produces a smaller output vector (averages/takes the maximum of each filter); this layer follows the same syntax as the flattening layer.

Listing 3-10. The architecture of an example convolutional encoder without pooling and only using convolutional layers with a vector-based bottleneck. For simplicity, the convolutional component has been reduced to two convolutional layers

```
encoder = Sequential()
encoder.add(L.Input((64, 64, 3)))

#convolutional component
encoder.add(L.Conv2D(8, (3,3)))
encoder.add(L.Conv2D(16, (3,3)))

#flattening and processing
encoder.add(L.Flatten())
encoder.add(L.Dense(256, activation='relu'))
encoder.add(L.Dense(32))
```

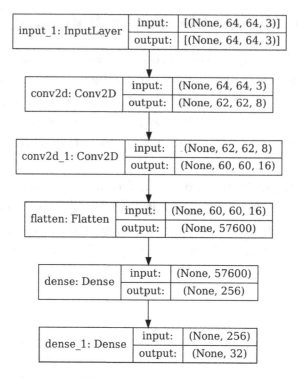

Figure 3-10. *The architecture of an example convolutional encoder without pooling and only using convolutional layers, using a vector bottleneck design*

Visualizing is especially helpful to help us understand transformations to the shape. We see that before flattening, the encoder had encoded the image into an image of shape (60, 60, 16), which was flattened into a vector with dimension 57,600. The output of the encoder is a vector of dimension 32.

We need this information to construct the decoder (Listing 3-11, Figure 3-11). The encoder takes in the encoded representation, which is a vector of dimension 32. It then uses Dense layers to increase its size up to the same size as the data right after the encoder had flattened it, which is (57600,). From there, we can reshape the vector into an image of shape (60, 60, 16) such that inverse convolutional operations can be applied to it.

Listing 3-11. The architecture of an example convolutional decoder without pooling and only using convolutional layers with a vector-based bottleneck

```
decoder = Sequential()
decoder.add(L.Input((32,)))
```

```
#processing and reshaping
decoder.add(L.Dense(256, activation='relu'))
decoder.add(L.Dense(57_600, activation='relu'))
decoder.add(L.Reshape((60, 60, 16)))

#applying transpose-convolutional layers
decoder.add(L.Conv2DTranspose(16, (3,3)))
decoder.add(L.Conv2DTranspose(8, (3,3)))
decoder.add(L.Conv2D(3, (1,1)))
```

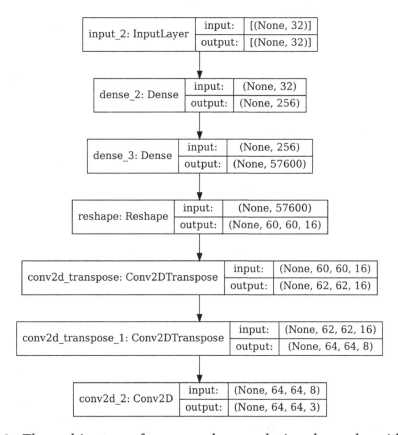

Figure 3-11. *The architecture of an example convolutional encoder without pooling and only using convolutional layers, using a vector bottleneck design*

Recall that to address the number of channels in the input image, in this case we put an extra convolutional layer with filter size (1,1) to maintain the image size but to collapse the number of channels.

139

Convolutional Autoencoder with Pooling and Padding

While we were technically successful in building a convolutional autoencoder in that the inputs and outputs were identical in shape, we failed to adhere to a fundamental principle of autoencoder design: the encoder should progressively decrease the size of the data. We need pooling in order to cut down on the image size quickly.

Convolutional neural networks are generally constructed in modules (Figure 3-12) of convolutional layer – convolutional layer – pooling layer (with the number of convolutional layers varying). These modules can be repeated over and over again. By following several convolutional layers of feature extraction with a pooling layer, these extracted features can be aggregated, and the key highlights can be passed onto the next convolutional layers for further feature extraction and processing. (You can find a more detailed discussion of module/cell-based architectural design in Chapter 6.)

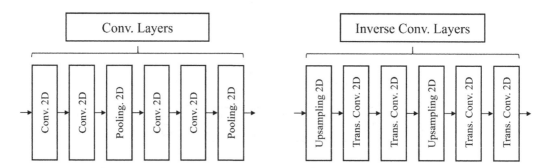

Figure 3-12. *Example repeated module/cell-based design in convolutional autoencoders*

However, when using convolutional layers in conjunction with pooling, we run into the problem of image sizes not divisible by the pooling factor. For instance, consider an image of shape (64, 64, 3). After a convolutional layer with filter size (2, 2) is applied to it, the image has a new shape of (63, 63, 3). If we want to apply pooling with size (2, 2) to it, we need to use padding in order to determine if the output of pooling will be (31, 31, 3) or (32, 32, 3). This is hardly a concern with standard convolutional neural networks. However, in autoencoders, we need to consider not only operations in the encoder but also the corresponding inverse "undoing" operations in the decoder. The upsampling layer has no padding layer. Thus, if we applied upsampling with size (2, 2) to an image with size (31, 31, 3), we would obtain an image of size (62, 62, 3); if we applied it to an image with size (32, 32, 3), we would obtain an image of size (64, 64, 3). In this case, there is no easy way in which we can obtain the original image size of (63, 63, 3).

You could attempt to work with specific padding and add padding layers manually, but it's a lot of work and difficult to manipulate and organize systematically.

To address this problem, one approach of building convolutional autoencoders with pooling and convolutional layers is to use padding='same' on all convolutional layers. This means that convolutional layers have no effect on the shape of the image – images are padded on the side before the convolution is performed such that the input and output images have identical shapes. The convolution is still changing the content of the image, but the image size remains constant. Removing the effect of convolutional layers significantly simplifies the management of image shape. Beyond this simplification, padding also allows for convolutions to process features on the edge of the image that might be passed over without padding by adding more buffer room such that edge features can be processed by the center of the kernel.

Let's build an encoder for an autoencoder with padding='same' for all convolutional layers (Listing 3-12).

Listing 3-12. The architecture of an example convolutional encoder with pooling and padding, using a vector-based bottleneck

```
encoder = Sequential()
encoder.add(L.Input((64, 64, 3)))

#convolutional component
for i in range(3):
    encoder.add(L.Conv2D(2**(i+3), (3,3),
                padding='same'))
    encoder.add(L.Conv2D(2**(i+3), (3,3),
                padding='same'))
    encoder.add(L.MaxPooling2D((2,2)))

#flattening and processing
encoder.add(L.Flatten())
encoder.add(L.Dense(256, activation='relu'))
encoder.add(L.Dense(32))
```

From visualizing (Figure 3-13), we can see that even though we prevented the convolutional layers from decreasing the size, with the addition of pooling, we were able to decrease the feature map resolution significantly. The flattened vector (before any processing) is 2048-dimensional – compare this to the 57600-dimensional post-flattening vector without pooling!

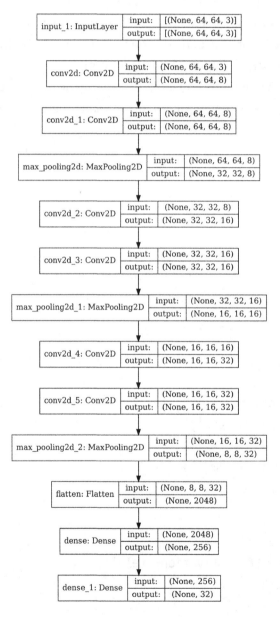

Figure 3-13. *Complete convolutional encoder with pooling and padding*

We can correspondingly construct the decoder (Listing 3-13, Figure 3-14). Like the convolutional layers in the encoder, the transpose convolutional layers in the decoder must use padding='same' to ensure symmetry.

Listing 3-13. The architecture of an example convolutional decoder with pooling and padding, using a vector-based bottleneck

```
decoder = Sequential()
decoder.add(L.Input((32,)))

#processing and reshaping
decoder.add(L.Dense(256, activation='relu'))
decoder.add(L.Dense(2048, activation='relu'))
decoder.add(L.Reshape((8, 8, 32)))

#applying transpose-convolutional layers
for i in range(3):
    decoder.add(L.UpSampling2D((2,2)))
    decoder.add(L.Conv2DTranspose(2**(3-i), (3,3),
                padding='same'))
    decoder.add(L.Conv2DTranspose(2**(3-i), (3,3),
                padding='same'))

#adding additional layer to collapse channels
decoder.add(L.Conv2D(3, (1,1)))
```

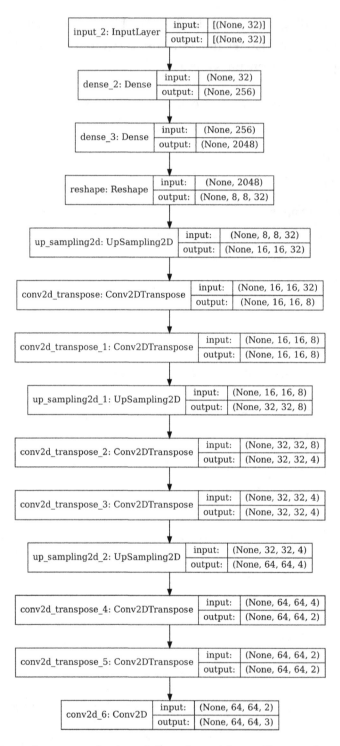

Figure 3-14. *Complete convolutional decoder with pooling and padding*

This method, of course, relies upon your input having a certain size. In this case, the input size must be a power of 2, since each pooling factor decreases the image's spatial dimensions by a factor of 2. You can insert a reshaping layer right after the input or reshape your dataset to accommodate this. The primary advantage of this method is that it makes organizing the symmetry of shape transformation much more simpler. If you know you want the shape of the encoded image right before flattening to be $(16, 16, x)^3$, for instance, and that you want to have three pooling layers with size $(2,2)$ and three pooling layers with size $(3,3)$, you can calculate the corresponding input shape to be of $16 \cdot 2^3 \cdot 3^3$ pixels in width and height.

Autoencoders for Other Data Forms

Using this logic, you can build autoencoders for data of all forms. For instance, you can use recurrent layers to encode and decode text forms of data, as long as an exact inverse decoding layer exists for every encoding layer you add. Although much of autoencoder work has been based around images, recent work is exploring the many applications of autoencoders for non-image-based data types. See the third case study in this chapter for an example of autoencoder applications in non-image data.

Autoencoder Applications

As we've seen, the concept of the autoencoder is relatively simple. Because of the need to keep the input and output the same size, however, we've seen that implementing autoencoder architectures for complex data forms can require a significant amount of forethought and pre-planning. The good news, though, is that implementing the autoencoder architecture is – in most autoencoders – the most time-intensive step. Once the autoencoder structure has been built (with the preferred compartmentalized design), you can easily adapt it for several applications to suit your purposes.

---

[3] The number of channels here is left arbitrary because pooling does not affect the depth of a tensor.

Using Autoencoders for Denoising

The purpose of denoising autoencoders is largely implied by its name: the "de" prefix in this context means "away" or "opposite," and thus denoising is to move "opposite of" or to remove noise. Denoising is simple to implement and can be used for many purposes.

Intuition and Theory

In a standard autoencoder, the model is trained to reconstruct whatever input it is given. A denoising autoencoder is the same, but the model must reconstruct a denoised version of a noisy input (Figure 3-15). Whereas the encoder in a standard autoencoder only needs to develop a representation of the input image that can be decoded with low reconstruction error (which can be very difficult as is), the encoder in a denoising autoencoder must also develop a representation that is robust to any noise. Denoising autoencoders can be applied to denoise messy signals, images, text, tabular data, and other forms of data.

Note   Denoising autoencoders are often abbreviated as DAE. You may notice that this is in conflict with the abbreviation of "deep autoencoder." Because the term "denoising autoencoders" is relatively more established and clearly defined than "deep autoencoder," when you see the abbreviation "DAE" in most contexts, it should be safe to assume that it refers to a denoising autoencoder. For the sake of clarity, in this book, we will favor not using the abbreviation "DAE"; if it is used, it will refer to the denoising autoencoder rather than the deep autoencoder.

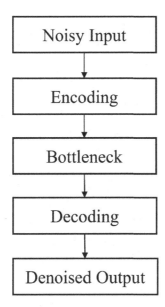

Figure 3-15. *Conceptual diagram of the components of a denoising autoencoder*

Of course, the concept of the denoising autoencoder assumes that your data is relatively free from noise in the first place. Because the denoising autoencoder relies upon the original data as the ground truth to perform denoising on the noisy version of that original data, if the original data is heavily noisy itself, the autoencoder will learn heavily arbitrary and noisy representation procedures. If the data you are using to train the denoising autoencoder has a high degree of noise, in many cases, there is little difference between using a denoising autoencoder and a standard autoencoder. In fact, using the former may be more damaging to your workflow because you may be operating under the assumption that the denoising autoencoder is learning meaningful representations of the data robust to noise when this is not true.

The noise to the input can be constructed through three methods (Figure 3-16): by inserting noise as a layer, into the dataset directly, or as a data generator. Which method to use depends on your problem type:

- *Apply noise as a layer*: Insert a random noise-adding layer directly after the input of the autoencoder such that the noise is applied to the input before any encoding and decoding is performed on it. The primary advantage of this method is that the model learns to be robust to multiple noisy versions of the same original image, since each time the model is passed through the model, different noise is applied. However, the noise-adding layer needs to be removed before the denoising autoencoder is used in application; it only serves as an artificial instrument during training, and in application, we expect the input to already be "naturally" noisy. When using this method, you can create the dataset like you would create it for an autoencoder – the input and outputs are the same.

- *Apply noise to the dataset directly*: Before training, apply noise to the dataset directly during its construction such that the data contains *x* as noisy data and *y* as the original data. The primary advantage of using this method is customizability: you can use any functions you would like to construct the dataset, since it is outside the scope of the neural network and therefore not subject to the restrictions of Keras and TensorFlow. You may want to add complex forms of noise that are not available as default layers or generators in Keras/TensorFlow. Moreover, there is no need to manipulate individual noise-adding layers of the autoencoder. However, you run the risk of overfitting (especially with small datasets and large architectures), because the autoencoder only sees one noisy form of each original input instance. Of course, you could manually produce several noisy forms of each instance, although it may take more time and be less efficient.

- *Apply noise through a data generator*: Keras/TensorFlow contains an `ImageDataGenerator` object that can perform a variety of augmentations and other forms of noise to the image, like adjusting the brightness, making small rotations and shears, or distorting the hue. Moreover, the image data generator is similar to the layer-based method in that the network is exposed to many different noisy representations of the input data – data is passed through the random generator in each feed-forward motion and distorted before it is formally processed by the network. The primary advantage of using the data generator is that you can apply forms of noise that are more natural or expected to occur in the data than with layers, which can implement more "artificial" forms of noise like adding Gaussian noise to the image. Moreover, there is no need to manipulate noise-adding layers after training the denoising autoencoder. However, image data generators are limited only to images, which means you will need to use another method for other forms of data.

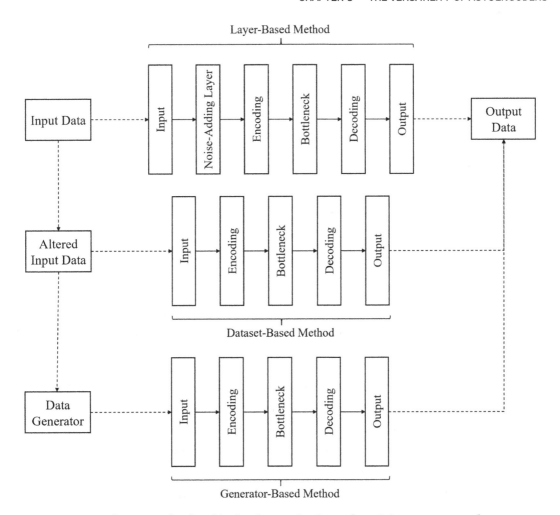

Figure 3-16. *Three methods of inducing noise in a denoising autoencoder*

Denoising autoencoders are primarily used, as expected, to denoise input data we expect will be noisy. You can insert the denoising autoencoder directly after the input such that the neural network performing some supervised task receives a denoised version of the input (Figure 3-17).

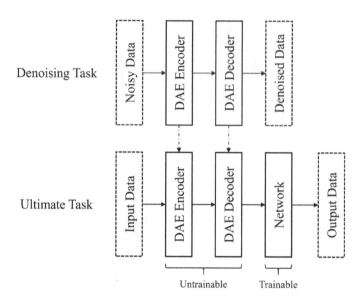

Figure 3-17. *One method of using denoising autoencoders to perform denoising on input data before the denoised input is processed. You can also unfreeze the denoising autoencoder to fine-tune afterward*

Consider a task in which a model must classify the sound of an audio input. For example, the model should be able to output a representation of the label "bird" if the audio input is of a bird song, or it should be able to output a representation of the label "lion" if the audio input is of a lion's roar. As one may imagine, a challenge in this dataset of real-life audio is that many of the audio inputs will contain overlapping sounds. For instance, if the audio clip of the bird song was from a park surrounded by the metropolitan bustle, the background would contain sounds of cars driving and people talking. To be successful, a model will need to be able to remove these sources of background noise and isolate the core, primary sound.

To do this, say you obtain another dataset of real-life sounds without background noise. You artificially add some form of noise that resembles the noise you would encounter in the noisy dataset. For instance, you can use audio processing libraries in Python to overlay a sound without background noise with smaller, dimmed down background noise. It is key for the success of the denoising autoencoder that the noise you artificially generate resembles the expected form of noise in the data. Adding Gaussian noise to images, for instance, may not help much as a denoising task unless the dataset you would like to denoise contains Gaussian noise. (You may find that adding some form of noise is better than adding no noise at all in terms of creating a self-supervised pretraining task, but this will be discussed more later.) You train a denoising autoencoder to reconstruct the original unnoisy sound signal from the noise-overlayed signal and then use the denoising autoencoder to denoise real-life sounds before they are classified.

Denoising can occur in other forms, though, too. For instance, you can denoise the entire dataset with the denoising autoencoder before training a model on the ultimate task instead of architecturally inserting the denoising autoencoder into the model operating on the ultimate task. This could be successful if you expect the "ultimate" model to be applied to relatively clean data in deployment, but know that the training data available is noisy. It's important to understand your particular problem well so you can successfully implement and manipulate your denoising autoencoder design.

Implementation

In our discussion of the implementation of denoising autoencoders, we will assume the autoencoder is being used for tabular data. The logic and syntax still can be applied to the use of denoising autoencoders for other data, like image or sequence data, with the necessary considerations for that particular data format.

Inducing Noise

As discussed, there are three practical methods of inducing noise.

One method is to insert a noising layer directly after the input (Listing 3-14, Figure 3-18). The most useful layer for this method will likely be the keras.layers.GaussianNoise layer, which adds Gaussian noise with a specified standard deviation. For a very heavy form of noise, you can also use layers like dropout or other modifications (although this is stretching the limits of what we would consider to strictly be a denoising autoencoder). You can also write your own custom layer for more complex noise-adding operations.

Listing 3-14. Layer-based method of inducing noise into a denoising autoencoder

```
ae_input = L.Input((784,), name='input')
ae_noising = L.GaussianNoise(0.1, name='noising')(ae_input)
ae_encoder = encoder(ae_noising)
ae_decoder = decoder(ae_encoder)
ae = Model(inputs = ae_input,
           outputs = ae_decoder)
```

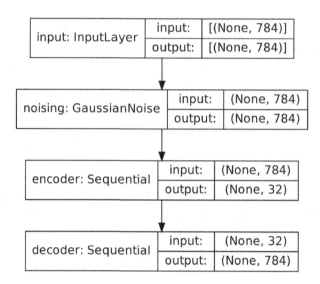

Figure 3-18. *Example implementation of inducing noise via layer method*

Another method of inducing noise is to apply it to the dataset directly, ideally during parsing when creating a TensorFlow dataset (Listing 3-15). When creating the parsing function, you can use a variety of TensorFlow image operations to induce random forms of noise. In this example, the unparsed TensorFlow dataset consists of a filename and a label for each instance (although you may find it helpful to arrange the unparsed dataset in another format if you know you will only use it for unsupervised learning).

In this example, we are using two helpful `tf.image.random_x` operations to induce random noise into the image, which vary brightness and JPEG quality noise. (The latter refers to a phenomenon in JPEG images in which visual artifacts distort the image due to the compression algorithm. This is especially helpful if your dataset consists of images in JPEG format or contains images transferred electronically that likely were once in JPEG format.) It should be noted that when specifying the range of noise – for instance, in JPEG quality, "80" and "100" in the example are the minimum and maximum JPEG quality – you should leave the possibility of cases in which no noise is applied to the image. This allows for the model not only to denoise a noisy image but also to recognize the presence of noise in the first place. We build this possibility by including the upper JPEG quality bound as 100%, in which the output of the `random_jpeg_quality` function is identical to the input. Functions like `random_brightness` require only for a maximum noise bound, and thus you can assume that there will be some examples with minimal levels of noise that satisfy this good practice.

Listing 3-15. Example function to pre-alter the dataset with noise using `.map()` on a TensorFlow dataset

```
def parse_file(filename, label):
    orig_img = do_preprocessing_stuff(filename)
    rand_img = tf.image.random_brightness(image, 0.01)
    rand_img = tf.image.random_jpeg_quality(image, 80, 100)
    return rand_img, orig_img
```

A third method to induce certain noise into image data is to use the Keras/ TensorFlow ImageDataGenerator (Listing 3-16). We begin by instantiating an ImageDataGenerator object with augmentation parameters for noise (the range to randomly shift brightness, small levels of shearing, etc.). To provide the image data generator object with data, use `.flow` or `.flow_from_x`. If using `.flow_from_directory`, be sure to set `class_mode='input'` so the image data generator does not assume labels and arrange the data generator as a classification problem.

Listing 3-16. Using the Image Data Generator method of inducing noise into an image dataset. Substitute augmentation_params for augmentation parameters. See Chapter 1 for a detailed discussion of Image Data Generator usage

```
from keras.preprocessing.image import ImageDataGenerator
idg = ImageDataGenerator(rotation_range=30,
                         width_shift_range=5,
                         ...)

idg_flow = idg.flow_from_directory(
    directory = '/directory/data',
    class_mode = 'input'
)
```

You can also use ImageDataGenerator with `class_mode='input'` as an alternative data source for autoencoders generally (not just for denoising) instead of using TensorFlow datasets. If you do decide to use image data generators for autoencoders, be sure to be careful with how you control your augmentation parameters for your particular purpose. If you are training a standard autoencoder, for instance, in which the input is identical to the ideal output, make sure to eliminate all sources of artificial noise by adjusting the augmentation parameters accordingly.

Using Denoising Autoencoders

If you are using the layer method of inducing noise into the denoising autoencoder, you will need to remove the noise-adding layer when using the denoising autoencoder in another application (Listing 3-17). Assume that the variable ae refers to the denoising autoencoder with a noise-adding layer, compartmentalized design, and proper naming. We can use the .get_layer(name) method to retrieve a layer or groups of layers by name and build them into a "final denoising autoencoder" model without the noise-adding layer. The example retrieves only the encoding and decoding layers, which store the weights that perform the key steps of encoding while denoising and decoding, thus eliminating the noise-adding layer in the final model. Here, you can see the many benefits that naming, compartmentalized design, and other organizational good practices can offer in manipulating the network structure.

Listing 3-17. Removing the noise-inducing layer in a denoising autoencoder

```
final_dae_input = L.Input((784,), name='input')
final_dae_encoder = ae.get_layer('encoder')(final_dae_input)
final_dae_decoder = ae.get_layer('decoder')(final_dae_encoder)
final_dae = Model(inputs=final_dae_input,
                  outputs=final_dae_decoder)
```

If you are dealing with a model that is not compartmentalized or that has many layers that need to be transferred, you can also refer to each layer by its index (Listing 3-18). Begin by defining the input layer and the layer after the input layer. In this case, the layer/group of layers we would like to follow the input layer in the final denoising autoencoder model is the encoder, which we can reference as ae.layers[2] because it is the third layer/component in the original denoising autoencoder architecture containing the noise-adding layer. From there, we can iteratively loop through the remaining layers and attach each one to the previous using Functional API syntax.

Listing 3-18. Alternate method of removing the noise-inducing layer in a denoising autoencoder

```
inp = L.Input((784,), name='input')
x = ae.layers[2](inp)
for layer in ae.layers[3:]:
    x = layer(x)
final_dae = Model(inputs=inp, outputs=x)
```

Keep in mind that there are complications with this sort of method when using nonlinear topologies.

By using any of these two methods, you can transfer the relevant weights into a final model without the intermediate layer. You will find this method of "surgically" removing an unwanted layer from a model helpful in other applications, especially in transfer learning. The resulting model (Figure 3-19) with the removed noise layer can be used for validation testing.

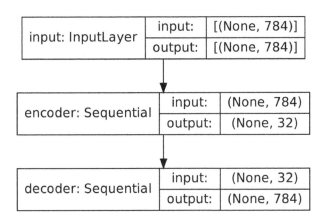

Figure 3-19. *The denoising autoencoder after removing the Gaussian noise layer*

Once you have obtained a "cleaned" autoencoder, you can insert it into another model as a preprocessing segment to denoise the input data (Figure 3-20, Listing 3-19).

Listing 3-19. Using a denoising autoencoder to denoise inputs before they are passed onto another model for further processing

```
process = Sequential(name='processing')
process.add(L.Input((784,)))
process.add(L.Dense(256, activation='relu'))
process.add(L.Dense(1, activation='sigmoid'))

inp = L.Input((784,), name='input')
denoising = ae(inp)
processing = process(denoising)
ult_model = Model(inputs=inp,
                  outputs=processing)
```

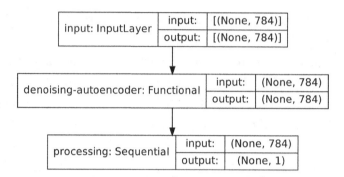

Figure 3-20. *Using the denoising autoencoder to decode the input before it is passed to the following layers for processing*

You can set your own layer freezing strategy (see Chapter 2) as necessary to suit your own purposes. For instance, you may decide to freeze the weights in the denoising autoencoder for the majority of training and then to unfreeze the entire network and perform some fine-tuning afterward.

Using Autoencoders for Pretraining

Another application of autoencoders is for pretraining. Recall that pretraining is used to provide "context" such that it develops certain skills or representations that allow it to succeed in performing its ultimate task. In Chapter 2, we discussed various pretraining methods and strategies. We will build upon this prior discussion to demonstrate when and how to use autoencoders in the context of pretraining. With both extensive knowledge of autoencoders and of pretraining at this point, you will find that the intuition and implementation of autoencoders for pretraining is quite straightforward.

Intuition

The use of autoencoders in pretraining falls under the category of self-supervised learning. You could think of autoencoders as the simplest form of self-supervised learning. Recall that in self-supervised learning, a model is trained on an altered dataset, which is constructed only on the input data, not the labels, of the task dataset. Some self-supervised learning tasks, for instance, involve predicting the degree by which an image was rotated or the degree of noise that was added to some set of data. In a standard autoencoder, however, no alterations to the data are needed beyond moving data instances into a dataset such that the input and output are the same for each instance.

Generally, when autoencoders are used for pretraining, the entire autoencoder is trained on the pretraining dataset, the encoder is extracted, more layers are appended to the encoder, and the newly formed model is trained on the task dataset (Figure 3-21). Because the encoder extracts important representations of the data from the unsupervised pretraining task, the hope of using the encoder for pretraining is that it begins with the basic ability to represent and encode key features of the input. The appended layers can then process the encoded key features into an output for the task dataset.

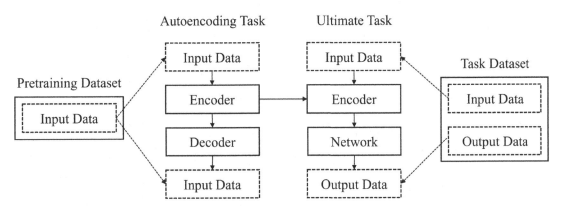

Figure 3-21. *Conceptual map of using autoencoders for pretraining. Input data is the same for autoencoders. "network" is used to refer to the layers appended to the encoder. This isn't entirely accurate, but it represents the idea that there is another "mini-network" that processes the output of the encoder to map whatever encoded representation comes out of the encoder into the output*

If you plan to use an autoencoder for pretraining, it's important you consider a key factor in how large to build the bottleneck (and correspondingly the widths of the encoder and the decoder components): the amount of information the layers appended after the encoder have to work with. If your bottleneck is large, the task may be trivial, and the network may not develop meaningful representations and processes for understanding the input. In this case, the processing layers after the encoder would receive a high number of features, but each feature would not contain much valuable information. On the other hand, if the bottleneck is too small, the model may develop meaningful representations, but the following processing layers may not have enough features to work with. This is a balancing process that takes experience and experimentation.

Like pretraining, it's best to freeze components strategically to make the most out of the weights in the encoder derived from pretraining (Figure 3-22). Like many pretraining tasks, generally it's good practice to freeze the pretrained component – in this case, the encoder – and train the following component(s) to better interpret the output of the encoder and then to unfreeze the pretrained component such that the entire network can be fine-tuned.

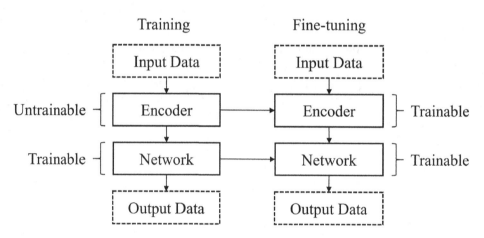

Figure 3-22. *How components of the autoencoder for pretraining are frozen or unfrozen throughout training and fine-tuning*

If you are using this method of pretraining with autoencoders with image data and you follow the design of flattening and reshaping the data around the bottleneck such that the bottleneck is a vector rather than an image, you will find that the structure of data transformation is especially clean. The encoder converts an image into a vector, and the following processing component processes that vector (which contains the encoded representation of the input) into the desired output. Thus, in this context, the encoder functions as the convolutional component and the following processing layers function as the fully connected component of an image-based deep learning model. This not only increases conceptual and organizational clarity but also allows you to further manipulate these sorts of autoencoder for pretraining designs for greater performance with the tools of transfer learning (see Chapter 2).

Note that while using autoencoders for pretraining, you can use a wide variety of autoencoder training structures beyond the standard autoencoder, in which the input is equivalent to the desired output.

For instance, denoising is a powerful self-supervised task that can be addressed with a denoising autoencoder (Figure 3-23).

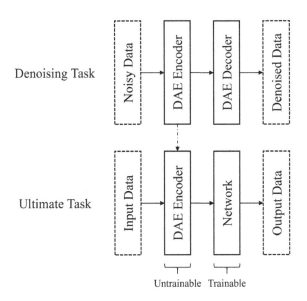

Figure 3-23. *Example of using non-standard autoencoder tasks for pretraining, like denoising autoencoder*

However, when you are choosing the self-supervised task to train your autoencoder, it's important to have a strong conceptual understanding of what that self-supervised task will accomplish. In this self-supervised context, the denoising autoencoder offers primarily a strong understanding of key features of the data, developed by identifying and correcting artificially inserted noise, and only secondarily actual denoising capabilities. This means that you do not need to be as careful about ensuring that the artificial noise resembles the true noise in the dataset when developing denoising autoencoders for pretraining. Of course, being conscious about the setup of the denoising autoencoder can allow you to maximize both the benefit of denoising (i.e., the encoder develops representations more or less robust to noise) and self-supervised learning (i.e., the encoder develops abstract representations of key features and ideas by learning to separate noisy artifacts from true meaningful objects).

Note   It should be noted that the concepts of denoising and self-supervised learning are not completely independent from one another. To properly denoise an input, a model must develop representations of key features and concepts within the input, which is the goal of self-supervised learning.

This sort of simplicity and conceptual ease in manipulating autoencoders for pretraining makes them exceptionally popular in modern deep learning design.

Implementation

Implementing autoencoders for pretraining is simple, given that you have used compartmentalized design. Recall the autoencoder structure built to demonstrate the construction of autoencoders for tabular data, which took in data with 784 features and compressed it into 32 neurons in the bottleneck layer before reconstructing the 784 features.

In the spirit of compartmentalized design, let's build a component that takes in the output of the encoder and processes it through a series of Dense layers to derive the output (in this case, for a binary classification problem, Listing 3-20). Note that this component takes in 32-dimensional data because the encoder outputs encoded representations that are 32-dimensional.

Listing 3-20. Building a sub-model to process the outputs of the encoded features

```
process = Sequential(name='processing')
process.add(L.Input((32,)))
process.add(L.Dense(16, activation='relu'))
process.add(L.Dense(10, activation='sigmoid'))
```

Even though we compiled the encoder and decoder models into the autoencoder, we can still reference the encoder and decoder individually, with their weights retained. The final model with only the pretrained encoder and the processing component can be built with the Functional API (Listing 3-21).

Listing 3-21. Using the processing sub-model with the encoder in an overarching model for a supervised task

```
inp = L.Input((784,), name='input')
encoding = encoder(inp)
processing = process(encoding)
ult_model = Model(inputs=inp,
                  outputs=processing)
```

Make sure to freeze layers as appropriate.

Using Autoencoders for Dimensionality Reduction

The concept of the autoencoder was initially presented as a method of dimensionality reduction. Because the application of dimensionality reduction is almost "built into" the design of the autoencoder, you will find that using autoencoders for dimensionality reduction is very simple to implement. However, there's still much to consider in performing dimensionality reduction for autoencoders; with the right design, autoencoders can offer a unique method of dimensionality reduction that is more powerful and versatile than other existing methods.

Intuition

Dimensionality reduction is generally performed as an unsupervised task, in which data must be represented with a smaller number of dimensions than it currently exists in. Many dimensionality reduction algorithms like Principal Component Analysis (PCA) and t-Stochastic Neighbor Embedding (t-SNE) attempt to project data into lower spaces according to certain mathematical articulations of what features of the data should be valued most in a reduction. PCA attempts to preserve the global variance, for instance, while t-SNE instead seeks to capture the local variance. Because different dimensionality reduction algorithms are built to prioritize the preservation of different features of the data, they are fundamentally different in character and thus are limited in their effectiveness on a wide variety of datasets.

However, in an autoencoder approach to dimensionality reduction, the autoencoder is trained to reconstruct the input after passing it through a bottleneck. After the autoencoder is fitted, the encoder is detached and the dimensionality reduction for any input can be obtained by passing it through the encoder and receiving its output (Figure 3-24).

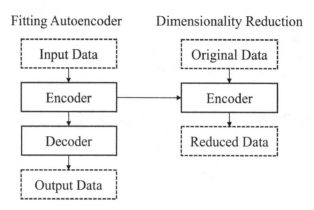

Figure 3-24. *Conceptual map of using autoencoders for dimensionality reduction*

Thus, autoencoders are distinctly different from other dimensionality reduction algorithms in two key ways: adaptability and articulation of valued features.

- *Adaptability*: Algorithms like PCA and t-SNE allow the user to adapt the algorithm to their dataset by manipulating a few parameters, but that quantity is vastly outnumbered by the adaptability of the autoencoder. Because autoencoders are more of a concept than an algorithm, they can be built with a much greater degree of adaptation to your particular problem. You can change the depth and width of the autoencoder structure, the activations in each layer, the loss function, regularization, and many other features of neural network architectural and training design to change how an autoencoder behaves in performing dimensionality reduction. This also means that using autoencoders for dimensionality reduction is likely to be successful only if you are aware of how different manipulations to the autoencoder structure translate to changes in its behavior *and* the behavior of a dimensionality reduction algorithm necessary to handle dimensionality reduction on your dataset.

- *Articulation of valued features*: Autoencoders prioritize certain features of the data and optimize the reduction of data in a way different from algorithms like PCA and t-SNE in character. Autoencoders attempt to minimize the reconstruction loss, while PCA and t-SNE attempt to maximize some relatively explicit mathematical formulation of what to prioritize, like the preservation of local or global structure (e.g., variance). These formulations attempt to capture what "information" entails in the context of dimensionality reduction. On the other hand, autoencoders do not seem, at least on the surface level, to have these prioritizations built explicitly into their design – they simply use whatever reduction allows for the most reliable reconstruction of the original input. Perhaps autoencoders are one of the most faithful representatives to capturing "information" in a broad, conceptual sense – rather than being tied toward any particular explicit assumptions of what constitutes a preservation of information (i.e., the *means* of preservation), it adopts whatever procedures and assumptions are necessary for the original item to be reconstructed (i.e., using whatever means are necessary to obtain optimal *ends* of preservation).

These two features can both be advantages and disadvantages. For instance, adaptability can be a curse rather than a tool if your dataset is too complex or difficult to understand. Moreover, increased adaptability does not necessarily suggest increased interpretability of adaptation; that is, while the autoencoder possesses a much wider range of possible behaviors, it is not necessarily simple to identify which changes to the architecture will correspond to certain changes (or absence of changes). Note that the autoencoder's articulation of valued features is determined by its loss function, which uses as one component the model prediction, which depends on the model architecture (among other modeling details). Finding your way through this chain of considerations is likely more arduous a task than adjusting relatively interpretable parameters on other dimensionality reduction methods.

Modern autoencoders for dimensionality reduction are most often used on very high-dimensional data since deep learning has evolved to be most successful on complex forms of data. More traditional algorithms like PCA developed for the reduction of lower-dimensional data are unlikely to be suited toward data like word embeddings in NLP-based models and high-resolution images. t-SNE is a popular choice for high-dimensional data, but primarily for the purpose of visualization. If you are looking to maximize the information richness of a dimensionality reduction and are willing to sacrifice some interpretability, autoencoders are generally the way to go.

Implementation

Using autoencoders for dimensionality reduction requires no further code from building and training the original autoencoder (see previous sections on building autoencoders for tabular and two-dimensional data). Assuming the autoencoder was built with compartmentalized design, you can simply call `encoder.predict(input_data)`, where encoder corresponds to the encoder architecture and `input_data` represents the data you would like encoded.

Using Autoencoders for Feature Generation

Feature generation and feature engineering is often thought of as a relic of classical machine learning, in which engineering a handful of new features could boost the performance of machine learning models. In deep learning applications involving relatively more complex forms of data, however, in many cases, using standard feature engineering methods like finding the row-wise mean of a group of columns or binning is unsuccessful or provides minimal improvement.

With autoencoders, however, we are able to perform feature generation for deep learning using an entity with the power and depth of deep learning methods.

Intuition

The encoding component of the autoencoder can be used to generate new features for a model to take in and process when performing a task. Because the encoder has learned to take in a standard input and compress it such that each feature in the encoded representation contains the most important information from the standard input, these encoded features can be exploited to aid the prediction of another model.

Functionally, the idea is almost identical to autoencoders for dimensionality reduction. However, using autoencoders for feature generation requires the additional step of generating new features and feeding the new features into the model.

We've seen earlier that a similar concept is employed in using autoencoders for pretraining, in which the encoder is detached from a trained autoencoder and inserted directly after the input of another network, such that the component(s) of the model after the encoder receive enriched, key features and representations of the input. However, the purpose of feature generation is to generate, or add, features rather than replacing them. Thus, when using autoencoders for feature generation, the encoder provides one set of encoded, information-rich features that are considered alongside the original set of features.

Consider, as an example, these two small datasets, consisting of the original features and the generated features of a hypothetical autoencoder fitted on the original features (Figure 3-25).

Original Features

F1	F2	F3	F4	F5
10.	5.8	5.7	3.2	4.6
2.8	6.	8.3	9.7	3.7
0.5	3.6	9.3	1.	7.2
0.5	4.5	6.	7.4	1.6

Generated Features

GF1	GF2
6.2	2.5
8.4	6.5
6.5	2.7
7.7	0.4

Figure 3-25. *Hypothetical datasets: original features and the generated features (produced by the output of an encoder in a trained autoencoder). The exact numbers in these tables are hypothetical (randomly generated)*

When using autoencoders for pretraining, the autoencoder is trained to reconstruct the original features and the encoder is transferred to the new model to provide a feature-extracting mechanism directly after the input (Figure 3-26). Thus, the remainder of the network, which performs much of the interpretation of these features, cannot access the original features directly. Rather, it understands the original features through the compression and "interpretation" of the encoder model.

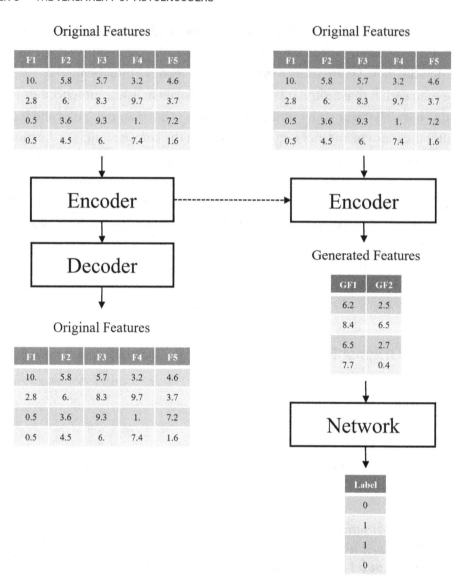

Figure 3-26. *Using only generated features from the encoder as inputs to the network*

However, a concern arises from this particular approach: even though the encoder from the autoencoder will almost always give the model a boost (if designed and trained properly), does the encoder impose a limit on the performance of the model by forcing it to take information strictly through the lenses of the autoencoder? Sometimes, a bit of fine-tuning after training (in which the encoder weights are unfrozen and the entire model is trained) is enough to resolve this concern.

Even fine-tuning may not be an adequate address of this concern, though, in several circumstances. If the autoencoder does not obtain a relatively high performance in reconstruction (i.e., middling, mediocre performance), forcing the model only to take in mediocre-level features could limit its performance. Alternatively, if the data is not of extremely high complexity, like tabular data or lower-resolution images, having an encoder compress the original inputs may be valuable, but not completely necessary. In many cases, models intended for less complex data benefit from processing both the original input and encoded features.

When using autoencoders for feature generation, the encoder's output is concatenated (or merged through some other mechanism) with the original input data, such that the remainder of the network can consider and process both sets of features (Figure 3-27). This method of feature generation can be thought of giving the remainder of the network the "best of both worlds" – access to the original, raw, unchanged data and a developed interpretation of that original data. Here, the network attempts to make the decision as to how each set of features is to be weighted and processed to optimize its performance.

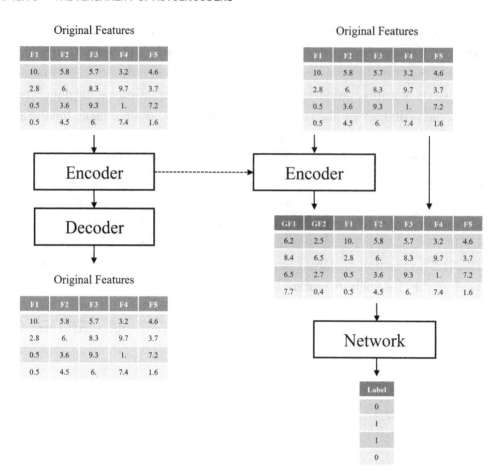

Figure 3-27. *Using both generated features from the encoder and original features as inputs to the network*

Architecturally, we can represent this with a nonlinear topology, in which the input layer feeds both into the encoder and the output of the encoder (Figure 3-28).

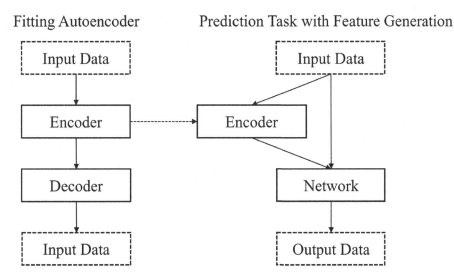

Figure 3-28. *Conceptual map of how fitted autoencoders can be used in a prediction task with feature generation in which both the original features and the generated features are inputted to the remaining network layers for processing*

This is similar in character to the layer-based method of inducing noise when training a variational autoencoder. You could pass data through the encoder and concatenate the encoded features to the original dataset and then feed the merged dataset into a standard network architecture. However, such an approach, in which changes to the data are performed outside of the neural network, is doing more work than necessary – it's much easier to use the Functional API to add the encoder than to get messy with predictions and data organization, especially if your dataset is on the larger side. Adding the encoder model directly into the new model automatically puts these relationships and data flows in place.

Implementation

Like many autoencoder applications, implementing autoencoders for feature generation is straightforward with the tools we've developed prior. Three lines of code using the Functional API allow us to define the input layer, the encoder, and the junction at which the original data and the generated features are merged (Listing 3-22).

Listing 3-22. Creating the feature generation component of the autoencoder, in which inputs are passed through the encoder and concatenated to those outputs

```
inp = L.Input((784,))
encoding = encoder(inp)
merge = L.Concatenate()([inp, encoding])
```

Afterward, we can add more Dense layers to process the merged features (Figure 3-29, Listing 3-23).

Listing 3-23. Processing the concatenated features

```
dense1 = L.Dense(256, activation='relu')(merge)
dense2 = L.Dense(64, activation='relu')(dense1)
output = L.Dense(1, activation='sigmoid')(dense2)
model = Model(inputs=inp, outputs=output)
```

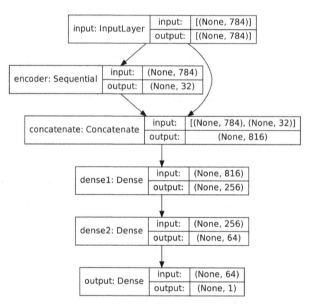

Figure 3-29. *Hypothetical map of the architecture of an autoencoder used for feature generation*

Be sure to set encoding.trainable = False, as the encoder's weights – the basis for the method by which it extracts core features and representations – should be frozen during training. The need for fine-tuning is less significant than if the autoencoder were used for pretraining.

You can also build more complex topologies to better take advantage of the encoder's encoded features by first processing the original features and the generated features independently before merging and following with further processing (Figure 3-30, Listing 3-24).

Listing 3-24. Processing the encoded representation and the original input separately before they are concatenated

```
inp = L.Input((784,))
encoding = encoder(inp)

# processing the input independently
p_inp_1 = L.Dense(256, activation='relu')(inp)
p_inp_2 = L.Dense(128, activation='relu')(p_inp_1)

# processing the encoder output independently
p_encoding_1 = L.Dense(32, activation='relu')(encoding)
p_encoding_2 = L.Dense(32, activation='relu')(p_encoding_1)

# merge and process
merge = L.Concatenate()([p_inp_2,
                         p_encoding_2])
dense1 = L.Dense(256, activation='relu')(merge)
dense2 = L.Dense(64, activation='relu')(dense1)
output = L.Dense(1, activation='sigmoid')(dense2)

# aggregate into model
model = Model(inputs=inp, outputs=output)

# freeze encoder weights
encoding.trainable = False
```

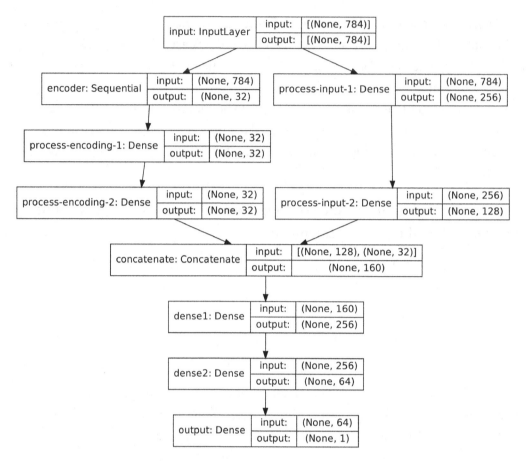

Figure 3-30. *Hypothetical map of the architecture of an autoencoder used for feature generation with further processing on the output of the encoder and the original features before concatenation*

Broadly, you can use these sorts of architectural manipulations to build all sorts of complex nonlinear topologies to take advantage of pretraining methods.

Using Variational Autoencoders for Data Generation

The variational autoencoder is one of the more modern conceptions of the autoencoder. It serves a relatively recent developing subfield in deep learning: data generation. While variational autoencoders are most often employed in the generation of images, they also have applications in generating language data and tabular data. Although image generation can be used to generate photorealistic images, more practically variational autoencoders are often used to generate more data to train another model on, which can be useful for small datasets. Because variational autoencoders heavily rely upon the notion of a latent space, they allow us to manipulate their output by traversing the latent space in certain ways. This allows it to offer more control and stability in the generated outputs than other data generation methods, like Generative Adversarial Networks (GANs).

Intuition

The goal of an autoencoder is to reconstruct the original input data as identically as possible. On the other hand, the goal of the variational autoencoder is to produce a similar image with reasonable variations – hence, the name "*variational*" *autoencoder*.

The fundamental idea behind generation using the variational autoencoder isn't too complex: not only individual points within the latent space (corresponding to data points in the dataset) are meaningfully arranged so that they can be decoded into their original form, but the latent *space* itself – consisting of all the space in and around existing data points – is mapped to by the encoder such that it contains relationships relevant to the dataset. Therefore, we can sample points from the latent space not occupied by existing encoded representations of data points from the dataset and decode them to generate a new data item that wasn't part of the original dataset.

For instance, consider data in two classes (perhaps images of dogs and cats or of the handwritten digits "0" and "1"). Of course, the autoencoder is not given these labels when trained, but it likely is able to perform separations and placement of these images such that these two classes are generally separated from one another (Figure 3-31). Thus, if we want an image resembling a dog, we can pass in an actual image of a dog from the dataset, make some changes to the corresponding location in the latent space to introduce variation, and decode the deviation/variation point to output a slightly different image of a dog.

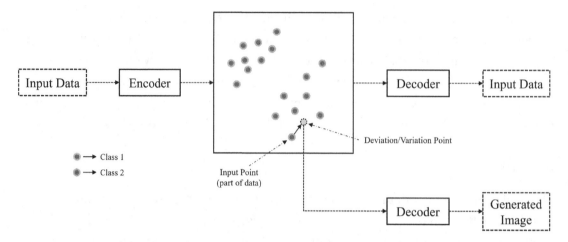

Figure 3-31. *Decoding regions of the latent space near the latent space point of a data point to generate variations of the data – data that is different, but that shares similar features. The box in the center represents the latent space – the points in the bottleneck region to which input data has been mapped to and which will be mapped back into a reconstructed version*

Alternatively, you could randomly select several points from the latent space and decode them to generate corresponding images (Figure 3-32). This method is more commonly used to generate a large and diverse array of data, since making small deviations or variations to an existing point still restricts the generated image to be relatively similar to the original image. By completely randomly selecting several points from within the latent space, you can produce a much more diverse generated dataset.

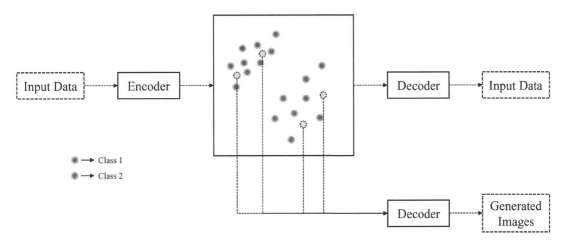

Figure 3-32. *Sampling randomly from the latent space to produce a wide, diverse dataset of generated data*

We can write the general process of generating data using an autoencoder at this point as follows:

1.   Train an autoencoder on a reconstruction task.

2.   Randomly select several points in the latent space, either by making changes to existing data points (ensuring that it is within a reasonable domain).

3.   Decode each of those points into the respective generated image.

While this general logic and intuition is valid, there is a problem with an assumption we have made: with standard autoencoders, the latent space is not continuous; rather, it's much more likely to be heavily discrete, separated into clusters. Continuity is unlikely to be helpful in the task of reconstructing inputs, because it magnifies the effect of small "slips" and "mistakes"; rather, having a discrete design allows for some minimum "guarantee" of success (Figure 3-33).

For instance, consider an autoencoder attempting to reconstruct images of the digits "0" and "1." Assume the latent space of an autoencoder is discrete, meaning that there were purposeful gaps separating clusters of images (in this case, images appearing to contain the digit "0" vs. images appearing to contain the digit "1"). Even if the encoder did not well encode an image of an image containing the digit "0" (it was not positioned in the optimal position in the latent space), it would still likely be within the cluster of images labeled "0," because there are large gaps in the latent space that separate off images containing "0" vs. containing "1." This means that the decoder would still reconstruct an image with a similar shape to "0" – even if it didn't capture the specifics of the image, it captures the main idea.

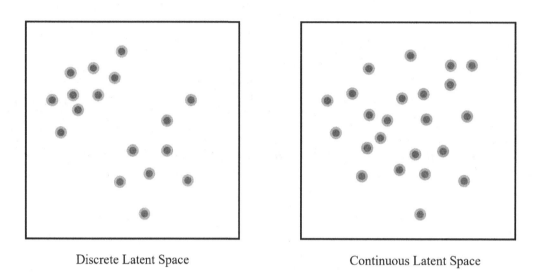

Discrete Latent Space   Continuous Latent Space

Figure 3-33. *A discrete vs. continuous latent space in autoencoders*

On the other hand, assume the latent space is continuous, meaning that there are no purposeful gaps separating clusters of images. If there is a similar deviation in the positioning of an encoded image containing the digit "0," it may be reconstructed by the decoder as "1" – there is no gap, or barrier, that separates the concepts of "0" and "1" from each other.

Thus, discreteness within autoencoders is a useful tool for autoencoders to improve the performance of their reconstruction by sectioning off main ideas. It has been empirically observed that successful autoencoders tend to produce discrete latent spaces.

However, this becomes a challenge for generating data, because when we randomly sample from the latent space, we assume that the space is continuous. What would happen, for instance, if we sampled a point in the latent space that happened to reside in one of the "gaps" between discrete clusters (Figure 3-34)?

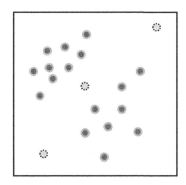

Figure 3-34. *Sampling from "gap" regions of the latent space that the decoder has not learned to decode leads to bizarre and wildly unrealistic inputs*

You may hypothesize that decoding a randomly sampled point in this "gap" would result in some sort of image that is "in between" both of the classes. However, the decoder has never been exposed to points in the latent space from that region and therefore does not have the experience and knowledge to interpret it. Therefore, the decoder will likely output some sort of gibberish that is in no way related to the original dataset.

Thus, variational autoencoders employ two key changes to force the autoencoder to develop continuous latent spaces, such that the latent space can both be sampled randomly (enabling a diverse generated dataset) and that the decoded latent space points will produce realistic/reasonable outputs (enabling a plausible generated dataset). Fundamentally, these changes force the autoencoder to learn continuous distributions rather than discrete locations in the latent space:

- The encoder learns the optimal mean and standard deviation of variables in the latent space, from which an encoded representation is sampled and passed on to the decoder for decoding.

- The variational autoencoder is optimized on a custom loss function: a blend of reconstruction loss (this is the cross-entropy, mean squared error, or any other standard autoencoder loss function used) and KL divergence to prevent the model from "cheating" (more on this "cheating" later).

After the input is passed through several processing layers to encode the data, it is separated into two vectors, representing a vector of means and a vector of standard deviations. Assuming the latent variables are drawn from a normal distribution, we can use the mean and standard deviation to randomly sample an encoded representation. This means that the same input can be encoded twice with different encoded representations. This encoded representation – now with some variation – is decoded and trained as an output (Figure 3-35).

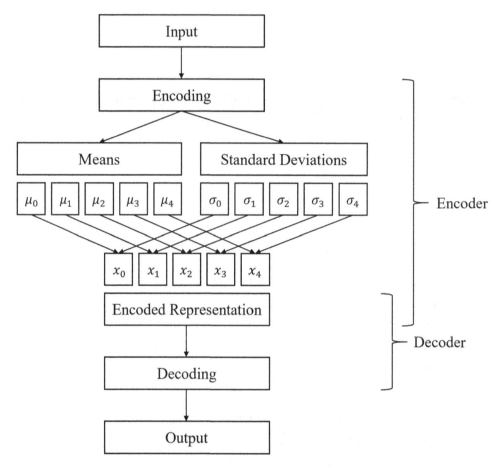

Figure 3-35. *Architecture of a variational autoencoder*

The architecture of the variational autoencoder exposes a certain beauty in deep learning: we can profoundly shape the neural network's thinking processes to our own desires with a few well-placed nudges. For instance, you may be wondering: what if certain latent variables aren't normally distributed? Over time, the network will adapt the encoding and decoding processes to accommodate the normal distribution because we use the normal distribution assumption to choose the encoded representation. If it did not adapt, it would not attain optimal performance. We can expect the variational autoencoder to learn this adaptation with high certainty because modern neural networks possess reliably high processing power. This, in turn, allows us to more freely impose (reasonable) expectations within models as a means to accomplish some end.

Similarly, even though we would like the branches after encoding to represent the means and standard deviations of the latent variables, we don't build any hands-on mechanisms before these two branches to manually tell the network what the mean or standard deviation is. Rather, we assume that each of the branches takes the role of the mean and the standard deviation of a distribution when generating the encoded representation (the expectation) and allow the network to adapt to that expectation. This sort of expectation-based design allows you to implement deep learning ideas faster and more easily.

With enough training, the autoencoder gradually develops an understanding of the probability distribution of each latent variable, allowing the latent space to be, by design, continuous (Figure 3-36). Instead of learning the absolute location of points, the autoencoder learns probabilistic "clouds" around each original input to gain experience with variations on the original input. This allows the autoencoder to interpolate in the latent space more smoothly.

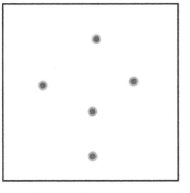 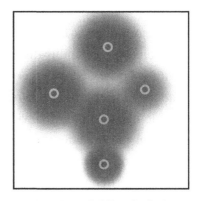

Learning Absolute Points
(Standard Autoencoders)

Learning Probability Distributions
(Variational Autoencoders)

Figure 3-36. *Standard autoencoders learn the absolute locations of data in the latent space, whereas variational autoencoders learn probability distributions*

However, there's still one way the network can cheat: by setting the branch representing the standard deviation to be arbitrarily small, the network can reduce the variation around each learned point in the latent space such that it is functionally identical to learning the location of absolute points (Figure 3-37). Since we know that autoencoders will tend toward building discrete spaces, variational autoencoders implement another mechanism to force the autoencoder to learn a continuous space: KL divergence.

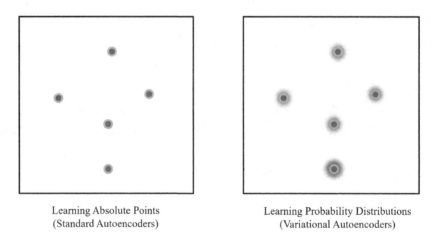

Learning Absolute Points
(Standard Autoencoders)

Learning Probability Distributions
(Variational Autoencoders)

Figure 3-37. *In practice, variational autoencoders can "cheat" and replicate the absolute point learning of standard autoencoders by forming clustering patterns and reducing the standard deviation*

KL divergence, or Kullback-Leibler divergence, measures the level of divergence, or difference, between two probability distributions (roughly speaking). For the purposes of variational autoencoders, KL divergence is minimized when the mean is 0 and the standard deviation is 1 and thus acts as a regularization term, punishing the network for clustering or reducing the standard deviation.

If the variational autoencoder is optimized only on KL divergence, it will tend to cluster all the points toward the center without any regard for decoding ability, because KL divergence is concerned with the distribution parameters outputted by the encoder, not the decoder's ability. Hence, like only using reconstruction loss, the latent space cannot be interpolated (Figure 3-38).

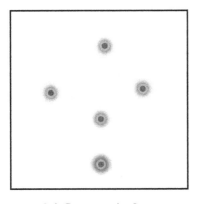 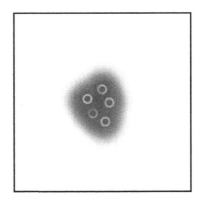

Only Reconstruction Loss   Only KL Loss

Figure 3-38. *Results of latent space representations when only using reconstruction loss vs. only using KL loss*

To address this problem, variational autoencoders are trained on the sum of reconstruction loss and KL loss. Thus, the variational autoencoder must simultaneously develop encoded representations that can be decoded into the original representation but will be punished if the representations make use of clustering, low standard deviations, and other marks of a discrete space. The variational autoencoder is a unique model in that its objective cannot be clearly articulated by any one existing loss function and thus is formulated as a combination of two functions.

Now, you can random uniformly sample the latent space within the domain of learned representations and produce both a diverse and realistic set of generated images. Moreover, our intuitive logic of interpolation applies: if you want to visualize what lies "in between" two images (e.g., a cat and a dog or the digit "1" and "9"), you can find the point in between the corresponding two points in the latent space and decode it.

Implementation

Implementing variational autoencoders is more difficult than implementing other autoencoder applications, because we need to access and manipulate components within the autoencoder itself (the latent space), whereas prior the only significant step was to change the input and output flow. Luckily, this serves as a useful exercise in the construction of more complex neural network structures in Keras.

Tip   Recall that in the Functional API, each layer is assigned to a variable. Although you should always be careful and specific with your variable naming, this is especially true with complex and nonlinear neural network architectures like the variational autoencoder. You'll need to keep track of the relationships between several layers across several components, so establish a naming convention and code organization that works best for you. If you make mistakes, you'll run into graph construction errors. Frequently use plot_model with show_shapes=True if you need guidance in debugging these errors to show what went wrong.

Let's begin by implementing the encoder. Prior, we used the Sequential model-building structure to construct the encoder and decoder, because the encoder and the decoder were not topologically nonlinear and we did not need to reference any intermediate layers. However, in variational autoencoders, both of these conditions are true; thus, we will build the encoder using the Functional API (Listing 3-25).

Listing 3-25. Implementing the first layers of the encoder for a variational autoencoder

```
enc_inputs = L.Input((784,), name='input')
enc_dense1 = L.Dense(256, activation='relu',
                     name='dense1')(enc_inputs)
enc_dense2 = L.Dense(128, activation='relu',
                     name='dense2')(enc_dense1)
```

After the inputs are processed by several Dense layers (or convolutional or any sort of appropriate layer, given your data type), the network splits into two branches, representing the mean and the standard deviation of the latent variable distribution (Listing 3-26). Note that in practice the network learns the log standard deviation rather than the raw standard deviation to prevent negative solutions. You may also see some approaches that use log variance, which is functionally no different from using standard deviation (variance is just standard deviation squared). Recall that with expectation-based design, we can expect the network to learn expectations as long as we set them (and the expectation is not too difficult to reach), so you can more or less direct the model to learn whatever attribute you would like it to.

Listing 3-26. Creating the branches to represent the mean and log standard deviations of the latent variable distributions

```
means = L.Dense(32, name='means')(enc_dense2)
log_stds = L.Dense(32, name='log-stds')(enc_dense2)
```

We've built two branches, representing the mean and log standard deviation of the latent variable distributions. In this case, our latent space has 32 dimensions (like a standard autoencoder with 32 bottleneck nodes), with one mean value and one log standard deviation value to describe the distribution. To attain the output of the encoder – the encoded representation of the input – we need to take in both the mean and the standard deviation, randomly sample from that distribution, and output that random sample (Listing 3-27). This sort of operation isn't built as a default layer in Keras/TensorFlow, so we're going to have to build it ourselves.

To build a custom layer, we can take advantage of the `keras.layers.Lambda()` method of constructing layers that perform operations. It can be used as follows.

Listing 3-27. Structure of creating a custom layer that performs TensorFlow operations on inputs using Lambda layers

```
def custom_func(args):
    param1, param2 = args
    perform_tf_operation(param1, param2)
    ...
    return result

custom_layer = L.Lambda(custom_func)([inp1, inp2])
```

In this example, `custom_layer` takes in the outputs of two previous layers. However, Keras delivers this data together in one argument variable containing all of the arguments to the custom function, which can be unpacked within the custom function. Additionally, all operations must be Keras/TensorFlow operations (see Keras backend for Keras' low-level operations). Note that you can use methods like TensorFlow's py_func (see Chapter 1) to convert Python functions into TensorFlow operations, if necessary.

The custom sampling layer for the variational autoencoder takes in the means and the log standard deviations. In the custom sampling function, we take in the argument variable and unpack it into a tensor containing the means and log standard deviations for each latent variable. Then, we generate a random, normally distributed tensor of the shape of the means and log_stds objects as eps, or epsilon. We can use the formula $\mu + e^{\ln(\sigma)} \cdot \epsilon$, which simplifies cleanly and intuitively to $\mu + \sigma \cdot \epsilon$ (Listing 3-28).

Listing 3-28. Creating a custom layer to sample encoded representations from the mean and log standard deviations of latent variables

```
def sampling(args):
    means, log_stds = args
    eps = tf.random.normal(shape=(tf.shape(means)[0], 32),
                           mean=0, stddev=0.15)
    return means + tf.exp(log_stds) * eps

x = L.Lambda(sampling, name='sampling')([means, log_stds])
```

The shape parameter used in creating the epsilon variable is expressed in this example as (tf.shape(means)[0], 32). Alternative methods include (tf.shape (log_stds)[0], 32) and (tf.shape(means)[0], tf.shape(means)[1]): all capture the same idea. In order for the actual sampling $(\mu + \sigma \cdot \epsilon)$ to be successful, the tensors for the mean, standard deviation/log standard deviation, and epsilon normally distributed randomness need to be the same size. Because this data is one-dimensional, we know that the input shapes to the sampling layer will be in the form of (batch_size, latent_space_size). We haven't specified what the batch size is, so we simply used tf.shape(mean)[0]. However, we know the latent space size – 32 dimensions. With more complex data types, it's important to understand the shape of the inputs to this sampling data you're dealing with.

It should also be noted that while the mean with which the epsilon tensor is generated should remain at 0, the standard deviation can be changed depending on how much variation you're willing to introduce. More variation could allow the model to better interpolate and produce more realistic and diverse results, but too much could risk introducing too much noise to possibly be plausibly decoded. On the other hand, less variation allows the decoder to better decode the image but may result in less exploration and smooth interpolation of "gaps" in the latent space.

We can now aggregate these layers into the autoencoder (Listing 3-29).

Listing 3-29. Aggregating the encoder model

```
encoder = keras.Model(inputs=enc_inputs,
                      outputs=[means, log_stds, x],
                      name='encoder')
```

Note that we output the means, log_stds, and x, whereas in the original design, the encoder only truly outputs x, the sampled encoded representation. Recall that the loss function for variational autoencoders is specialized: a hybrid of reconstruction loss and KL divergence. In order to calculate KL divergence loss, the loss function requires access to the means and standard deviations. Therefore, in order to provide this access, we will list means and log_stds as outputs to the encoder model. While we won't use them in the decoder, this allows our loss function to work. The encoder architecture is as follows (Figure 3-39).

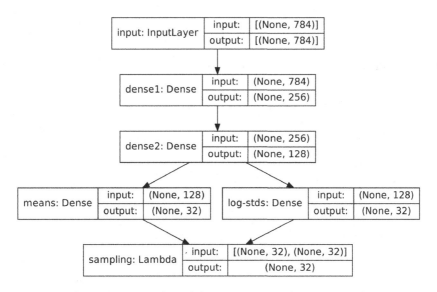

Figure 3-39. *Building the encoder of the variational autoencoder*

Building the decoder is relatively straightforward: like the standard autoencoder, it takes in an encoded representation (in this case, a 32-dimensional vector) and progressively decodes it (Listing 3-30).

Listing 3-30. Creating and aggregating the decoder model

```
dec_inputs = L.Input((32,), name='input')
dec_dense1 = L.Dense(128, activation='relu',
                        name='dense1')(dec_inputs)
dec_dense2 = L.Dense(256, activation='relu',
                        name='dense2')(dec_dense1)
output = L.Dense(784, activation='sigmoid',
                     name='output')(dec_dense2)
decoder = keras.Model(inputs=dec_inputs,
                        outputs=output,
                        name='decoder')
```

We can join these two components together into a variational autoencoder model (Listing 3-31).

Listing 3-31. Aggregating the encoder and decoder sub-models into an overarching variational autoencoder model

```
vae_inputs = enc_inputs
encoded = encoder(vae_inputs)
decoded = decoder(encoded[2])
vae = keras.Model(inputs=vae_inputs,
                    outputs=decoded,
                    name='vae')
```

This method of combining components together into a model is like the method that has been used for standard autoencoders with two key differences. Firstly, the input to the variational autoencoder is not a unique keras.layers.Input or keras.Input input mechanism, but the input to the encoder (enc_inputs). This is because of the design of the variational autoencoder's loss function: because the model must have access to the output of the encoder, the encoder's input must also be on the same "level" as the variational autoencoder itself. Note that this is a valid operation because we're not making any actual architectural changes (i.e., rerouting layers or adding multiple branches to one layer), only simplifying a technically redundant (but perhaps organizationally helpful) process of passing data through two input layers. Secondly, the decoder does not take in the entire encoder output, but instead indexes it (decoded = decoder(encoded[2])). Recall that for the KL divergence component of the loss function to have access to the distribution parameters, the encoder outputs the tensors containing the mean and log standard deviation, as well as the actual encoded representation. The decoder only takes in the encoded representation, which is the third output of the encoder, and is thus indexed accordingly.

All that is left before fitting is to construct the loss function. Recall that our loss function is a combination of the reconstruction loss (we'll use binary cross-entropy in this case) and KL divergence loss. Creating the reconstruction loss is easy because it's already implemented in Keras (Listing 3-32).

Listing 3-32. Building the reconstruction loss component of the variational autoencoder's custom loss function

```
from keras.losses import binary_crossentropy
reconst_loss = binary_crossentropy(vae_inputs, decoded)
```

Building KL divergence is more difficult. Although KL divergence is already implemented in Keras/TensorFlow, in our case we need to build a custom simplified KL divergence that does not take in predictions and ground truth labels, but instead means and standard deviations, measuring how far the distribution is from a distribution with mean 0 and standard deviation 1 (Listing 3-33).

Listing 3-33. Building the KL divergence loss component of the variational autoencoder's custom loss function

```
kl_loss = 1 + log_stds - tf.square(means) - tf.exp(log_stds)
kl_loss = tf.square(tf.reduce_sum(kl_loss, axis=-1))
```

Note that `tf.reduce_sum` allows you to sum the losses for each item across a specified axis; this is necessary to deal with batching, in which there is more than one data point per batch.

The final variational autoencoder loss can be expressed as a sum of the reconstruction loss and the KL divergence loss (Listing 3-34). `tf.reduce_mean` averages the losses for each of the data points to produce the final loss value.

Listing 3-34. Combining reconstruction and KL divergence loss into the VAE loss

```
vae_loss = tf.reduce_mean(reconst_loss + kl_loss)
```

We need to make one more accommodation for the variational autoencoder's unique loss function: usually, the loss function is passed into the model's `.compile()` function. However, we can't do this in this case because the inputs to our loss function are not predictions and ground truth labels that the Keras data flow assumes loss functions take in when passed through compiling: instead, it takes in the outputs of specific layers. Because of this, we use `vae.add_loss(vae_loss)` to attach the loss function to the variational autoencoder.

The model can then be compiled (without the loss function, which has already been attached) and fitted on the data. Since there are so many crucial parameters that determine the behavior of the variational autoencoder (architecture size, sampling standard deviation, etc.), it's likely that you'll need to tweak the model parameters a few times before you reach relatively optimal performance.

We've taken several measures to adapt our model structure and implementation to work with the unique nature of the variational autoencoder's loss function. While this book – and many tutorials – will present the building of these sorts of architectures in a linear format, the actual process of constructing these models is anything but sequential. The only way to truly know if the unique situation you are working with requires a certain workaround or needs to be rewritten or re-expressed somehow is to experiment. In this case, it is likely helpful to first write down the basic architectures of the encoder and the decoder, implement the custom loss function, and then test and experiment which changes to the architecture and/or loss function are needed to make it all work, consulting documentation and online forums as necessary.

Case Studies

In these case studies, we will discuss the usage of autoencoders for a wide variety of applications in recent research. Some can be accessed via existing implementations, whereas others may be out of the reach of our current implementation skills – nevertheless, these case studies provide rich and interesting examples of the versatility of autoencoders and put forth empirical evidence of their relevance in modern deep learning.

Autoencoders for Pretraining Case Study: TabNet

Deep learning is quite famously seldom used for tabular data. A common critique is that deep learning is "overkill" for simpler tabular datasets, in which the massive fitting power of a neural network overfits on a dataset without developing any meaningful representations. Often, tree-based machine learning models and frameworks like XGBoost, LightGBM, and CatBoost are employed instead to handle tabular data.

However, especially with larger tabular datasets, deep learning has a lot to offer. Its complexity allows it to use tabular data in a multimodal fashion (considering tabular data alongside image, text, and other data), engineer features optimally on its own, and use tabular data in deep learning domains like generative modeling and semi-supervised learning (see Chapter 7).

In the 2020 paper "TabNet: Attentive Interpretable Tabular Learning,"[4] Sercan O. Arik and Tomas Pfister propose a deep learning architecture for tabular data named TabNet. Arik and Pfister highlight three key aspects of TabNet's architecture and design:

- *End-to-end design*: TabNet does not need feature processing and other preprocessing methods required when using tabular data for other approaches.

- *Sequential attention*: Attention is a mechanism that is commonly used in language models to allow the model to develop and understand relationships between various entities in a sequence. In TabNet, sequential attention allows the model to choose, at each step, which features it should consider. This, moreover, allows for greater interpretability of the model's decision-making processes.

- *Unsupervised pretraining*: TabNet is pretrained using an autoencoder-based architecture to predict masked column values in a tabular fashion. This leads to a significant performance and improvement. TabNet demonstrated, for the first time, large improvement from self-supervised pretraining on tabular data.

---

[4] Arik, S. Ö. and Pfister, T. (2021). TabNet: Attentive Interpretable Tabular Learning. *Proceedings of the AAAI Conference on Artificial Intelligence*, 35(8), 6679–6687. Retrieved from https://ojs.aaai.org/index.php/AAAI/article/view/16826. Code link (Keras/TensorFlow implementation of TabNet with minor changes): https://github.com/titu1994/tf-TabNet.

For the purposes of this case study, the unsupervised pretraining via autoencoder structure is the focus.

Using an encoder and a decoder, TabNet was trained via self-supervised learning to reconstruct masked column values (Figure 3-40). After pretraining is completed, the encoder is detached and altered to be fitted for a supervised task. By pretraining such that the decoder and reconstruct missing inputs based on the TabNet encoder's encoded representations, the TabNet encoder develops efficient processes for extracting and aggregating key information given existing information. Moreover, it must develop a strong understanding of the relationships between each of the tabular dataset's columns. These representations and knowledge allow TabNet to more easily be adapted toward some sort of decision-making, in which the information-rich encoded representations need only to be further interpreted and adapted for the particular classification task.

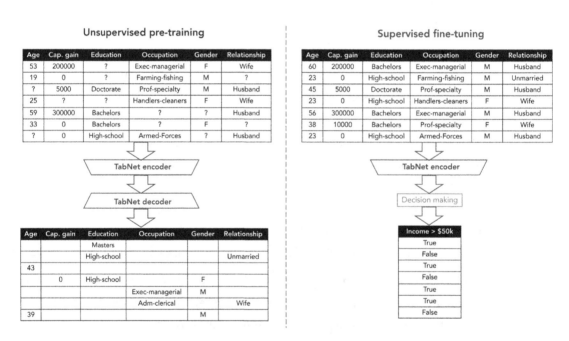

Figure 3-40. *TabNet method of unsupervised pretraining before supervised fine-tuning. Produced by Arik and Pfister*

When tested on a wide range of problems, TabNet performs the same or better than standard decision tree-based approaches to deep learning (Table 3-2). The Forest Cover Type dataset, introduced in 2017, is a classification task; TabNet outperforms XGBoost,

LightGBM, and CatBoost by a significant margin. It also has improved performance over the AutoML Tables automated search framework without using as thorough a parameter search as AutoML Tables required.

Table 3-2. *Performance on Forest Cover Type dataset*

Model	Test Accuracy (%)
XGBoost	89.34
LightGBM	89.28
CatBoost	85.14
AutoML Tables	94.95
TabNet	96.99

The Sarcos dataset is a problem relating to the dynamics of an anthropomorphic robot arm, in which 7 joint positions, 7 joint velocities, and 7 joint accelerations (a 21-dimensional input space) must be mapped to 7 joint torques. A TabNet model with only 6300 parameters has a lower MSE on this dataset than any other tested non-TabNet model (Table 3-3). A large TabNet model with 1.75 million parameters achieves an astoundingly low test error far from that of other models.

Table 3-3. *Performance on the Sarcos dataset*

Model	Test MSE	Model Size
Random forest	2.39	16.7 K
Stochastic decision tree	2.11	28 K
MLP	2.13	0.14 M
Adaptive neural tree	1.23	0.60 M
Gradient boosted tree	1.44	0.99 M
Small TabNet	1.25	6.3 K
Medium TabNet	0.28	0.59 M
Large TabNet	0.14	1.75 M

Moreover, experiments comparing the performance of TabNet trained with and without pretraining show that pretraining significantly improves performance, especially in cases in which an unlabeled dataset is much larger than the labeled dataset (Table 3-4). This makes sense – self-supervised pretraining allows the model to take advantage of unlabeled data that a purely supervised training process would not be able to exploit. Pretraining both increases test accuracy and increases the speed of convergence.

Table 3-4. *Performance on the Higgs Boson dataset with the medium-sized TabNet model. Similar improvements with pretraining are observed with other databases*

Training Dataset Size	Test Accuracy (%)	
	Without Pretraining	With Pretraining
1 K	57.47 ± 1.78	61.37 ± 0.88
10 K	66.66 ± 0.88	68.06 ± 0.39
100 K	72.92 ± 0.21	73.19 ± 0.15

To work with the TabNet model, install the tabnet library with pip install tabnet. Two relevant ready-to-go TabNet models are implemented: tabnet.TabNetClassifier and tabnet.TabNetRegressor, for classification and regression tasks, respectively (Listing 3-35). These models can be compiled and fitted like standard Keras models.

There are several key features in the TabNet classifier model to be aware of: feature_ columns accepts a list or tuple of feature columns to indicate the number of features but can instead be specified via num_features; num_classes indicates the number of output classes; feature_dim determines the dimensionality of the hidden representation in the feature transformation block; and output_dim determines the dimensionality of the outputs of each decision step.

Listing 3-35. Instantiating a TabNet classifier from the tabnet library

```
from tabnet import TabNetClassifier
model = TabNetClassifier(feature_columns=None,
                         num_classes=10,
                         num_features=784,
                         feature_dim=32,
                         output_dim=16)
```

```
model.compile(optimizer='adam',
              loss='sparse_categorical_crossentropy',
              metrics=['accuracy'])
model.fit(x_train, y_train, epochs=100)
```

It is important to fit the model with a high initial learning rate that gradually decays and a large batch size anywhere between 1% and 10% of the training dataset size, memory permitting. See the documentation for information on syntax for more specific modeling.

Denoising Autoencoders Case Study: Chinese Spelling Checker

English, like many other European languages, relies upon the piecing together of letter-like entities sequentially to form words. Misspelling in English can be addressed relatively well with simple rule-based methods, in which a rulebook maps common misspelled words into their correct spellings. Deep learning models, however, are able to consider the context of the surrounding text to make more informed decisions about what word the user intended to use, as well as fix grammar mistakes and other language errors.

However, Chinese "words" are characters rather than strings of letters. Mistakes in Chinese are commonly the result of two primary errors: visual similarity (two characters share similar visual features) and phonological similarity (two characters are pronounced similarly or identically). Because Chinese and English have different structures and therefore different requirements for correcting spelling, many of the deep learning approaches adapted for English and other European languages cannot be readily transferable to Chinese.

In "FASPell: A Fast, Adaptable, Simple, Powerful Chinese Spell Checker Based on DAE-Decoder Paradigm," Yuzhong Hong, Xianguo Yu, Neng He, Nan Liu, and Junhui Liu[5] propose a deep learning approach to correcting Chinese spelling based upon the denoising autoencoder architecture.

---

[5] Yuzhong Hong, Xianguo Yu, Neng He, Nan Liu, and Junhui Liu, "FASPell: A Fast, Adaptable, Simple, Powerful Chinese Spell Checker Based on DAE-Decoder Paradigm," 2019. Paper link: https://aclanthology.org/D19-5522.pdf. Code link: https://github.com/iqiyi/FASPell.

The approach consists of two components: a masked language model and a confidence-similarity decoder. The masked language model takes in sample text and outputs several candidates for replacement. The confidence-similarity decoder selects replacements from these candidates to form the completed, spell-checked sentence.

The masked language model, which functions as the denoising autoencoder (parsing "noise" – the spelling errors – from the input), uses a BERT-like structure. In models like BERT, pretraining is often performed by randomly masking certain words and training the model to predict the masked word. However, the errors induced by using random masks are not representative of the real-world errors seen in Chinese spelling. The pretrained masked language model in FASPell is fine-tuned on a composite dataset constructed with three processes:

- *Standard masking*: Some fraction of the dataset uses the same random masking as BERT: some character(s) is masked with a [MASK] token, and the model predicts which character is masked based on the context. This is performed on texts that have no errors.

- *Error self-masking*: Errors in the text are "masked" as themselves. The target label corresponds to the correct character. The model must identify that there is an error and perform a correction to obtain perfect performance.

- *Non-error self-masking*: Correct characters in the text are also "masked" as themselves, and the target label is that character. This is to prevent overfitting. In this case, the model must identify that there is no error and allow it to pass through the model unchanged.

While the masked language model provides the basis for the spelling correction, on its own is a weak Chinese spell checker – the addition of the confidence-similarity decoder to improve performance.

The decoder attempts to process and understand the denoised candidates generated by the masked language model. The decoder takes in both the confidence of the masked language model in each character and the similarity between characters. The similarity between two characters is calculated as a combination of their visual and phonological similarity.

The logic of comparing the masked language model's confidence and the character similarity is as follows (Figure 3-41). Say that the masked language model proposes a character replacement that has a high confidence and is very similar to the original erroneous character. The high confidence indicates that the masked language model is sure that the correction is correct. Because the original erroneous character is highly similar to the correction, we are more confident that the correction is correct because it is more likely for a mistake to be more than less similar to the correct character.

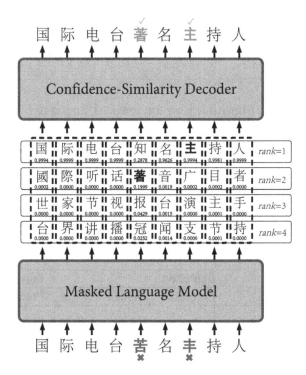

Figure 3-41. *Architecture and example data flow throughout the FASPell model, including the masked language model and the confidence-similarity decoder. Produced by Hong et al*

On the other hand, if the masked language model has a high confidence in a correction, but the correction and the original character have little similarity, these two signals must be closely compared. If the masked language model's confidence is high enough as to outweigh the low similarity between the original character and the correction, the correction is accepted. For each character in the sentence, the correction that maximizes the model confidence and the similarity with the original character is accepted and inserted into the final output.

The FASPell design achieves state-of-the-art performance in both error detection and correction (Table 3-5), outperforming prior Chinese spelling correction systems proposed by Wang et al. (2018) and Yeh et al. (2013).

Table 3-5. *Performance of FASPell in various spelling tasks, measured with three different metrics*

Model	Detection			Correction		
	Precision	Recall	F1	Precision	Recall	F1
Wang et al.	54.0	69.3	60.7	Not reported	Not reported	52.1
Yeh et al.	Not reported	Not reported	Not reported	70.3	62.5	66.2
FASPell	76.2	63.2	69.1	73.1	60.5	66.2

Moreover, removing the fine-tuning on the masked language model and removing the confidence-similarity decoder both generally significantly decrease performance, demonstrating the important role both elements play in adapting the traditional denoising autoencoder-style masked language model for the unique challenges of Chinese text (Table 3-6).

Table 3-6. *Performance of versions of FASPell in various spelling tasks, measured with three different metrics*

Model	Detection			Correction		
	Precision	Recall	F1	Precision	Recall	F1
FASPell	76.2	63.2	69.1	73.1	60.5	66.2
FASPell without fine-tuning	75.5	40.9	53.0	73.2	39.6	51.4
FASPell without confidence-similarity decoder	42.3	41.1	41.6	32.2	31.3	31.8
FASPell without fine-tuning or confidence-similarity decoder	65.2	47.8	55.2	48.4	35.4	40.9

FASPell shows the importance of adapting autoencoders toward your particular task. By changing the content and method by which the denoising autoencoder was trained and adding a confidence-similarity decoder to incorporate knowledge of character similarity into understanding the denoising autoencoder's output, FASPell was able to address a key shortcoming in standard pretrained autoencoder designs.

Variational Autoencoders Case Study: Text Generation

Earlier, we looked at variational autoencoders for image data. Recall that variational autoencoders are especially valuable because we can control the generated data by making changes to the sampled location in the latent space. This allows for fine control over attributes like the text's sentiment and writing style that is not possible with other generation methods.

Variational autoencoders generally work with the right design well on images. However, applying variational autoencoders for text generation, in which convolutional layers are replaced with recurrent layers, is more difficult. Recall that the loss for variational autoencoders is a sum of the reconstruction loss and KL divergence. KL divergence acts as a regularizing term that can be optimized by clustering the latent variable distributions closer together.

When training variational autoencoders for natural language tasks with recurrent based structures (Figure 3-42), the KL divergence term of the loss function tends to *collapse* or drop to 0. This means that all the latent variable distributions are essentially all on top of each other, clustered as closely together as possible. Because these overlapping distributions are very difficult to decode and make sense of, the decoder that generates the text completely ignores these latent representations. Instead, because RNN (recurrent neural network)-based decoders have significant modeling power, the decoder can produce text that scores a relatively low reconstruction error without relying on encoded representations provided by the decoder.

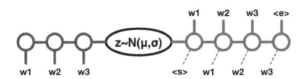

Figure 3-42. *The traditional LSTM-based variational autoencoder for text generation, which suffers from KL divergence term collapse*

This results in repetitive and unoriginal outputs that "cheat" the loss function, like these sample generated tweets.

```
@userid @userid @userid @userid @userid
@userid thanks for the follow
@userid @userid @userid @userid @userid @userid @userid @userid @userid
@userid @userid thanks for the follow
```

In "A Hybrid Convolutional Variational Autoencoder for Text Generation," Stanislau Semeniuta, Aliaksei Severyn, and Erhardt Barth[6] propose a variational autoencoder-based structure to address this problem. Using their approach, the generated output is both more realistic and diverse:

```
@userid All the best!!
@userid you should come to my house tomorrow
I wanna go to the gym and I want to go to the beach
@userid and it's a great place
@userid I hope you're feeling better
```

The architecture includes a hybrid of convolutional and recurrent layers; the encoder is composed completely of convolutional layers, whereas the decoder combines deconvolutions (like transpose convolutions) and a traditional RNN. Deconvolutional layers were chosen instead of recurrent layers for two key reasons: deconvolutional layers have more efficient GPU implementations, and feed-forward architectures are easier to optimize than recurrent ones, due to a more consistent and smaller number of steps required.

By using standard convolutional and deconvolutional layers in a variational autoencoder format, key features from the sequence can be learned and mapped to a continuous latent space (Figure 3-43). The sampled encoded representation can be decoded via a series of deconvolutions. The output of this convolutional variational autoencoder is passed through a conventional LSTM network for further processing (Figure 3-44). The recurrent component is specialized to consider dependencies between entities of various texts by considering sequence. This alleviates the burden of this task from convolutional/deconvolutional layers, which are not built to perform it successfully.

---

[6] Stanislau Semeniuta and Aliaksei Severyn and Erhardt Barth, "A Hybrid Convolutional Variational Autoencoder for Text Generation," 2017. Paper link: https://arxiv.org/pdf/1702.02390.pdf. Code link: https://github.com/ryokamoi/hybrid_textvae/blob/master/model/decoder.py (no-Keras TensorFlow implementation).

198

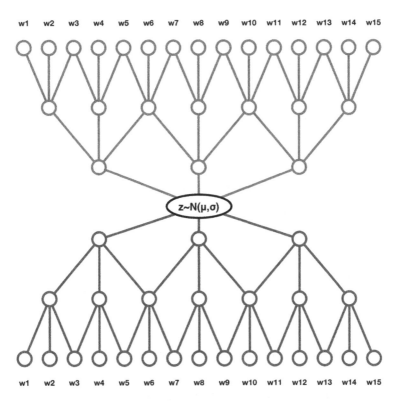

Figure 3-43. *The feed-forward component of the variational autoencoder. Inputs enter through the bottom and are processed through convolutions and deconvolutions to produce a generated set of features*

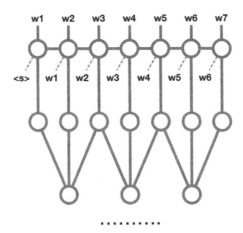

Figure 3-44. *The LSTM component of the hybrid approach, in which the generated features are further decoded by an LSTM-style model into text to capture complex sequence-based dependencies between entities of a sentence*

This architecture allows for meaningful text generation without employing overpowering stacks of recurrent layers. However, three other mechanisms are required to fully address the KL divergence collapse problem:

- *Input dropout*: A dropout layer is applied to the input such that the input, even before encoding, is already corrupted in a random fashion.

- *KL term annealing*: By gradually increasing the weight of the KL divergence term in the loss function across time, the neural network transitions from being a standard autoencoder into a variational autoencoder.

- *Auxiliary reconstruction term*: A specialized term is added to the loss, which can be optimized only by meaningful reconstruction and encoding. This exacerbates the importance of the reconstruction component of the loss function.

Using a convolutional recurrent hybrid variational autoencoder with appropriate modifications to the optimization strategy yields a much more successful model that is both able to produce diverse and realistic outputs. By linearly interpolating across two points in the latent space, we're able to transition from one tweet to a completely different tweet:

```
@userid I do that too. I have to wait a lot of people too
@userid I do not know about it and then I can find a contest
@userid I am so excited for this summer I hope you are well
@userid i don't know what to do in the morning i love you so much for the
shoutout
@userid i don't know what to do if you don't mind follow me i love you so much xx
@userid it would be awesome to hear it's a great place to see you around
the weekends
@userid it would be awesome to hear I'm so excited for the summer I'm going
to see them
```

Semeniuta, Severyn, and Barth's work shows how being creative with combinations of layers and transgressing prepackaged boundaries for what certain model types should look like, in addition to well-pointed, specific mechanisms to target particular challenges, can lead to significantly greater performance.

Key Points

In this chapter, we discussed the intuition and implementation of autoencoders for tabular and image data, as well as for five key autoencoder applications: denoising via denoising autoencoders, pretraining, dimensionality reduction, feature generation, and data generation via variational autoencoders.

- Autoencoders encode and decode data in the process developing important representations of the dataset and efficient processes for extracting key information and features from data. Autoencoders rely upon the context of the dataset to support their efficient representations. Therefore, while they are not good universal compression algorithms, they have many applications that exploit the autoencoder's unique ability to learn complex relationships and representations of data in an unsupervised fashion.

- A key trade-off in autoencoder design is the size of the bottleneck: a bottleneck that is too small may make any encoding task too difficult to be practical, whereas a bottleneck that is too large can be solved trivially without developing meaningful representations.

- Use compartmentalized design when implementing autoencoders and other complex neural network structures by defining models as a series of relationships between sub-models. This makes implementing autoencoder applications significantly easier and is also conceptually more clear to mentally understand and manipulate.

- To build an autoencoder for tabular data, simply stack Dense layers with a progressively decreasing number of neurons in the encoder and a progressively increasing number of neurons in the decoder. To build an autoencoder for image data, use convolutional layers and pooling layers in the encoder, and transpose convolutional layers and upsampling layers in the decoder. Convolutional autoencoders often flatten and reshape in the region surrounding the bottleneck such that the output of the encoder (the encoded representation) is a vector. Use `padding='same'` in convolutional layers for easy manipulation of the shape size. Autoencoders can also be built for other data types, as long as there exists an inverse decoding layer for every encoding layer.

- Denoising autoencoders are autoencoders trained to reconstruct the original data from a corrupted, noisy input. If you are using denoising autoencoders to denoise dataset, ensure that the form of noise is realistic and exists within the dataset. When implementing denoising autoencoders, noise can be induced either as a layer, applied to the dataset, or as an Image Data Generator (in the case of images).

- When using autoencoders for pretraining, the autoencoder is first trained on unsupervised data to extract key ideas and representations. Then, the encoder is detached and custom processing layers are added after it to alter the model structure such that it can perform a supervised task. The encoder weights are usually frozen such that the following layers can be trained to interpret the output of the encoded representations; this can be followed by some fine-tuning to optimize performance.

- Autoencoders are different from other dimensionality reduction algorithms in two key ways: higher adaptability and a model-centric articulation of valued features. These can be implemented simply by having the encoder predict encoded representations for certain data inputs.

- Using autoencoders for feature generation is like using them for dimensionality reduction, but the encoded features are concatenated to the original dataset rather than replacing the original dataset. This can be implemented with a nonlinear architecture, in which the input is both passed through the encoder and concatenated to the output of the encoder; this enriched data can then be passed through a series of processing layers to be interpreted and used to produce an output.

- Variational autoencoders allow for the generation of new data by choosing locations in the latent space and decoding them. However, because standard autoencoders tend to produce discrete spaces that cannot be sampled randomly, variational autoencoders implement two key changes to force development of a continuous latent space: learning the *probability distributions* of latent variables rather than the absolute position and adding KL divergence to the loss function to provide regularization in the formation of the latent space. Variational autoencoders rely upon relatively more low-level operations but can be accomplished with good organization and active experimentation.

Autoencoders are extraordinarily versatile structures that you can work into almost any deep learning design. Keep them in mind when you approach a new deep learning problem and need somewhere to start – autoencoders provide a good starting point for developing knowledge and building important representations.

In the next chapter, we'll discuss model compression and the exciting implications it holds for the future of deep learning.

CHAPTER 4

Model Compression for Practical Deployment

It is my ambition to say in 10 sentences what everyone else says in a book.

—Friedrich Nietzsche, Philosopher and Writer[1]

Over the course of deep learning's flurried development in these recent decades, model compression has relatively recently become of prominent importance. Make no mistake – model compression methods have existed and have been documented for decades, but the focus for much of deep learning's recent evolution was on expanding and increasing the size of deep learning models to increase their predictive power. Many modern convolutional networks today contain hundreds of millions of parameters, and Natural Language Processing models have reached hundreds of *billions* of parameters (and counting).

While these massive architectures push forward the boundaries of what deep learning can do, often their availability and feasibility are restricted to the realm of research laboratories and other high-powered departments within organizations that have the hardware and computational power to support such large operations. *Model compression* is concerned with pushing forward the "cost" of the model while retaining its performance as much as possible. Because model compression aims primarily at maximizing efficiency rather than performance, it is key to transferring deep learning advances from the research laboratory to practical applications like satellites and mobile phones.

---

[1] Nietzsche, Friedrich, "The Twilight of the Idols: Or, How to Philosophise with the Hammer," *Project Gutenberg*, 1911, www.gutenberg.org/files/52263/52263-h/52263-h.htm.

© Andre Ye 2022
A. Ye, *Modern Deep Learning Design and Application Development*,
https://doi.org/10.1007/978-1-4842-7413-2_4

Model compression is often a missing chapter in many deep learning guides, but it's important to remember that deep learning models are increasingly being used in practical applications that impose limits on how wildly one can design the deep learning model. By studying model compression, you can ground the art of deep learning design in a practical framework for deployment and beyond.

Introduction to Model Compression

When we perform model compression, we attempt to decrease the "cost" a model incurs while maintaining its performance as much as possible. The term "cost" used here is deliberately vague because it encompasses many attributes. The most immediate cost of storing and manipulating a neural network is the number of parameters it holds. If a neural network contains hundreds of billions of parameters, it will take more storage than a network that contains tens of thousands of parameters. It will be unlikely that applications with lower storage capability, like mobile phones, can even feasibly store and run models on the larger end of modern deep learning designs. However, there are many other factors – all related to one another – that factor into the cost of running a deep learning model:

Note   Since model compression is largely a concern of deployment, we'll use the corresponding language: "server side" and "client side." Roughly speaking, for the purposes of this book, "server side" refers to the computations performed on the servers servicing the client, whereas "client side" refers to the computations performed on the client's local resources.

- *Latency*: The latency of a deep learning model is the time taken to process one unit of data. Latency usually concerns the time it takes for a deployed deep learning model to make a prediction. For instance, if you're using a deep learning algorithm to recommend search results or other items, a high latency means slow results. Slow results turn away users. Latency is usually correlated with the size of the model, since a larger model requires more time. However, the latency of a model can also be complicated by other factors, like the complexity of a computation (say a certain layer does not have many parameters but performs a complex, intensive computation or a heavily nonlinear topology).

- *Server-side computation and power cost*: Computation is money! In many cases, the deep learning model is stored on the server end, which is continually calculating predictions and sending those predictions to the client. If your model is computationally expensive, it will incur heavy literal costs on the server side.

- *Privacy*: This is an abstract but increasingly important factor in today's technological landscape. Services that follow the preceding model of sending all user information to a centralized server for predictions (for instance, your video browsing history sent to a central server to recommend new videos, which are sent back to and displayed on your device) are increasingly subject to privacy concerns, since all user information is being stored at one point in a centralized location. New, distributed systems are increasingly being used (e.g., federated learning), in which a version of the model is sent to each user's individual device and yields predictions for the user's data without ever sending the user's data to a centralized location. Of course, this requires the model to be sufficiently small to operate reasonably on a user's device. Thus, a large model that cannot be used in distributed deployment can be considered to incur the cost of lack of privacy (Figure 4-1).

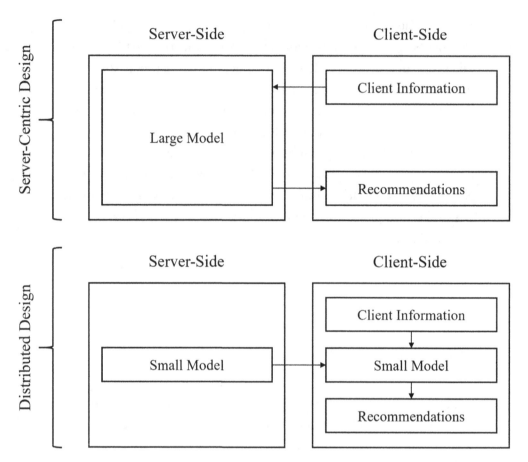

Figure 4-1. *Privacy requires a small model. While it may not be completely quantifiable, it's an important aspect of a model's cost*

These are all factors of a model's cost that must be considered during deployment, alongside the actual performance of the model. A model that incurs a low cost but performs poorly cannot be deployed any more in practical applications than a model that performs well but incurs a high cost. Research into neural networks has demonstrated that neural networks contain a certain amount of redundancy – additional space that is not needed at all for the particular problem. This makes sense: a small set of architectural designs can accommodate the vast majority of deep learning problems, but not all deep learning problems are the same in difficulty, and thus we should not expect each problem to "use" each architecture to the same amount. Removing the redundancy comes at no or negligible cost to the performance. Past this redundancy, though, we face a trade-off between performance and cost. As we decrease the cost a model incurs, we also decrease the performance of the model (Figure 4-2).

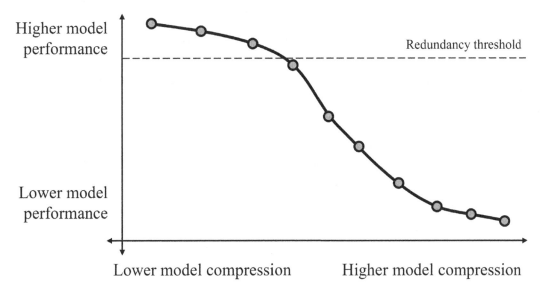

Figure 4-2. *A hypothetical relationship between model performance and model compression, with a redundancy threshold – the threshold before continual model compression leads to a dramatic decrease in model performance – marked. The actual relationship may vary by context and model compression type*

Which combination is optimal depends on your particular task and resource availability. Sometimes, performance is not a high priority compared to cost attributes. For instance, consider a deep learning model on a mobile application tasked with recommending apps or other items to open – here, it is much less important that the model be perfect than it is for the model to not consume much of the mobile phone's storage and computation. It is likely that a user would be unsatisfied net-wise if such an application consumed much of their phone's resources, even if such an application performed well. (In fact, deep learning may not even be the right approach in this situation – a simpler machine learning or statistical model may suffice.) On the other hand, consider a deep learning model built into a medical device designed to quickly diagnose and recommend medical action. It is likely more important for the model to be perfectly accurate than it is to be a few seconds faster.

Model compression is fascinating because it demonstrates the true breadth of problem-solving required to fully tackle a deep learning problem. Deep learning is not concerned only with improving model performance by the metrics but also about developing practical deep learning that can be used in real-life applications.

It also is valuable in advancing theoretical understandings of deep learning because it forces us to ask key questions about the nature of neural networks and deep learning:

if model compression can so effectively remove much of the information from a network with a marginal decrease in performance, what purpose did the original component of the network that was compressed "away" serve in the first place? Is neural network training fundamentally a process of *improving* solutions by tuning weights or a process of *discovery* – finding a good solution in a sea of bad solutions? How fundamentally robust are networks to small variations? We'll explore these questions along with a discussion on their benefits to practical deployment.

In this chapter, we will discuss three key deep learning model compression algorithms: pruning, quantization, and weight clustering. Other deep learning model compression/downsizing methods exist – notably, Neural Architecture Search (NAS). However, it is not discussed in this chapter because it is more appropriate in the next chapter.

Pruning

When you think of the number of parameters in a neural network, you likely associate parameters with the connections between nodes of each layer in a fully connected Dense network or perhaps the filter values in a convolutional neural network. When you call `model.summary()` and see the eight- or nine-figure parameter counts, you may ask: are all of those parameters necessary to the prediction problem?

As mentioned earlier, you can reasonably expect from simple reasoning that it is very likely you did not build a neural network architecture with the "perfect" number of parameters to succeed at its task, and thus if the network doesn't underperform, it likely is using more parameters than it truly needs. Pruning is a direct way to address these "superfluous" parameters by explicitly removing them. Its success in all sorts of architectures, moreover, poses important questions in theoretical deep learning.

Pruning Theory and Intuition

Imagine that you want to cut down on comforts in your living space – you think that there might be more than you need to work and that keeping all the comforts is increasing your cost of living beyond what it could be. You've made changes in your living space with an eye toward minimally reducing the impact on your ability to work – you still keep your computer, good Wi-Fi, steady electricity. However, you've cut down on comforts that may aid your work but that are fundamentally auxiliary by removing or donating items like a television subscription, a nice couch, or tickets to the local symphony.

This change in living space shouldn't theoretically impact your work, provided you are reasonably resilient – your mental facilities have not been explicitly damaged in a way that would impair your ability to perform the functions of work (perhaps it affects your comfort, but that's not a factor in this discussion). Meanwhile, you've managed to cut down on general costs of living.

However, you're a bit disoriented: you reach instinctively for a lamp in the corner that has been removed. You find yourself disappointed that you can't watch unlimited television anymore. Making these changes to this space requires *reorienting* yourself to it. It takes a few hours (or even days) of exploration and orientation to acclimate toward these new changes. Once you have completed acclimating, you should be ready to work in this newly modified space just as well as you did in your prior living space.

It should be noted, though, that if the disparity in your prior living space and your current living space is too large, you may never be able to recover – for instance, removing your exercise equipment, all sources of entertainment, and other items very close to but still not directly impacting your work. If you cut further into your living costs by cutting down on running water and electricity, your work ability would be directly impaired.

Let's rewind: you step back and take a look around your living space and decide that it has too many unnecessary comforts, and you want to cut these comforts down. You could take away all the comforts at once, but you decide instead that the immediate, absolute difference might be too stark for you to handle. Instead, you decide to embark on an *iterative* journey, in which you remove one or two things of the least significance every week, and stop whenever you feel that removing any more items would result in damage to your core working facilities. This way, you have time to acclimate to a series of small differences.

The logic of cutting down comforts in living spaces applies in parallel to the logic of pruning – it provides a useful intuitive model when feeling out how to perform pruning campaigns.

Pruning was initially conceived in Yann LeCun's 1990 work in "Optimal Brain Damage" – not all parameters contribute significantly to the output, so those parameters can be pruned away in the most optimal form of "brain" (neural network) damage.

Pruning is generally performed after the network has been substantially trained, such that evaluation of parameter importance is meaningful and not based only on random initialization or values in the early stages of training. In order to determine which neural network entities (nodes, connections, layers, etc.) contribute the most or least significantly to the output, each entity must be evaluated according to some importance criterion. The least important entities are removed (Figure 4-3). In practice, removal simply means setting a parameter to zero, which is much cheaper to store.

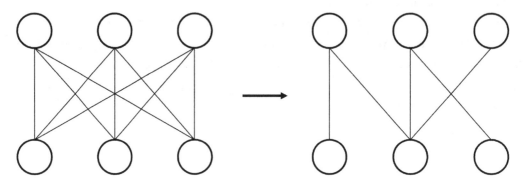

Figure 4-3. *Visualization of unstructured pruning*

This action of pruning away entire connections, nodes, and other neural network entities can be thought of as a form of architecture modification. For optimal performance, the model needs to be reoriented toward its new architecture via fine-tuning. This fine-tuning can be performed simply by training the new architecture on more data.

Thus, pruning follows the following general iterative process (Figure 4-4).

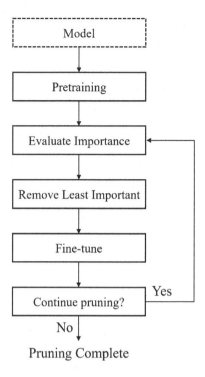

Figure 4-4. *Pruning process*

In this sense, you can think of pruning as a "directed" dropout, in which connections are not randomly dropped but instead pruned by the importance criterion. Because pruning is directed rather than random, a more extreme percentage of weights can be pruned with reasonable performance compared to the reasonable percentage of weights dropped in dropout.

When implementing pruning, you can specify the initial percent of parameters to be pruned and the end percent of parameters to be pruned; TensorFlow will figure out how much to remove at each step for you.

There have been many proposed methods of evaluating the importance of a parameter:

- *Magnitude pruning*: This is the simplest form of pruning: a weight is considered more important if it has a higher weight. Smaller weights are less significant and can be pruned with little effect on the model performance, given enough fine-tuning. Although many more complex methods have been proposed, they have usually failed to achieve performance of a significant margin higher than that of magnitude pruning. We will be using magnitude-based pruning methods in this book.

213

- *Filter pruning*: Pruning convolutional neural networks requires additional considerations, since pruning a filter requires removing all the following input channels that don't exist anymore. Using a magnitude-based pruning approach (average weight value in a filter) works well in convolutional networks.

- *Least effect on performance*: A more sophisticated compression method is to choose weights or other network entities that reduce the neural network cost change the most.

The operation of pruning only connections is known as unstructured pruning. Unstructured pruning can lead to sparse matrices, which can in many cases lead to computational difficulty and inefficiency. By pruning other larger neural network entities, you may be able to achieve even better compression at the cost of decreased precision (and potentially lower performance).

- *Pruning neurons*: Take the average of a neuron's incoming and outgoing weights and use a magnitude-based method to entirely remove redundant neurons. Other more sophisticated criteria can be used to prune entire neurons. These allow for groups of weights to be more quickly removed, which can be helpful in large architectures.

- *Pruning blocks*: Block-sparse formats store blocks contiguously in memory to reduce irregular memory access. Pruning entire memory blocks is similar to pruning neurons as clumps of network parts but is more mindful of performance and energy efficiency in hardware.

- *Pruning layers*: Layers can be pruned via a rule-based method – for instance, every third layer is pruned during training such that the model is slowly shrunk during training but adapts and compresses the information. The importance of a layer can also be determined by other analyses determining its impact on the model's output.

Each neural network and each task require a differently ambitious pruning campaign; some neural networks are built relatively lightweight already, and further pruning could severely damage the network's key processing facilities. On the other hand, a large network trained on a simple task may not need the vast majority of its parameters.

Using pruning, 90% to 95% of the network's parameters can reliably be pruned away with little damage to performance.

Pruning Implementation

To implement pruning (as well as other model compression methods), we're going to require the help of other libraries. The TensorFlow Model Optimization library works with Keras/TensorFlow models but is installed separately (`pip install tensorflow-model-optimization`). It should be noted that the TensorFlow Model Optimization library is relatively recent and therefore less developed than larger libraries; you may encounter a comparatively smaller forum community to address warnings and errors. However, the TensorFlow Model Optimization documentation is well written and contains additional examples, which can be consulted if necessary. We will also need the `os`, `zipfile`, and `tempfile` libraries (should be included in Python by default), which allow us to understand the cost of running a deep learning model.

Although TensorFlow Model Optimization significantly aids in the code required to implement pruning, it involves several steps and needs to be approached methodically. Moreover, note that because intensive work in pruning and model compression broadly has been relatively recent, at the time of this book's writing, TensorFlow Model Optimization does not support a wide array of pruning criterion and scheduling. However, its current offerings should satisfy the pruning needs of most compression needs.

Setting Up Data and Benchmark Model

For the purposes of this section, we'll train (and prune) a feed-forward model on a tabular version of MNIST data for the sake of simplicity. The logic applies to other more complex architectures, though, like convolutional or recurrent neural networks.

You can directly load the MNIST data from `keras.datasets` and make necessary adjustments with numpy and `keras.utils` (Listing 4-1).

Listing 4-1. Loading MNIST data

```
# import keras
import keras

# load mnist data
mnist = keras.datasets.mnist
(x_train, y_train), (x_test, y_test) = mnist.load_data()

# reshape from image data (28,28) into flat data (784,)
x_train = x_train.reshape((len(x_train), 28*28))
x_test = x_test.reshape((len(x_test), 28*28))

# one-hot encode labels
y_train = keras.utils.to_categorical(y_train)
y_test = keras.utils.to_categorical(y_test)
```

The MNIST dataset is notably a relatively simpler dataset, but for our benchmark model, we'll build a deliberately redundant model with many more neurons and layers than are needed. This model (Listing 4-2) will contain ten hidden layers, with groups of two layers containing the same successively decreasing power of 2 (from 512 to 32), such that the number of neurons in the hidden layer goes 512-512-256-256-128-128-....

Listing 4-2. Constructing a simple and redundant baseline model

```
# import layers
import keras.layers as L
# construct Sequential model
model = keras.Sequential()

# construct Input
model.add(L.Input((784,)))

# construct processing layers
for i in list(range(5,10))[::-1]:
    model.add(L.Dense(2**i, activation='relu'))
    model.add(L.Dense(2**i, activation='relu'))

# construct output layer
model.add(L.Dense(10, activation='softmax'))
```

We can correspondingly compile and fit the model with appropriate parameters (Listing 4-3).

Listing 4-3. Compiling and fitting baseline model

```
model.compile(optimizer='adam',
              loss='categorical_crossentropy',
              metrics=['accuracy'])
model.fit(x_train, y_train, epochs=15)
```

Creating Cost Metrics

As mentioned in the introduction, model compression is a trade-off. In order to understand the benefits of compression, we will need to create a few cost metrics for comparison to better understand factors like storage space, parameter count, and latency. These metrics can be applied not only to pruned models but compressed models in general; they serve as a North Star in navigating the compression trade-off.

Storage Size

To get the size of the file used to store the compressed model, we'll need to follow the following process:

1. Create a temporary file to store the model weights in.

2. Store the model weights in the created temporary file.

3. Create a temporary file to store the zipped model weight files in.

4. Obtain and return the size of the zipped file.

Let's begin by importing necessary libraries (Listing 4-4) – `zipfile` provides zipping functionality, `tempfile` allows for creating temporary files, and `os` allows for obtaining the size of a particular file.

Listing 4-4. Importing necessary libraries for storage size

```
import zipfile as zf, tempfile, os
```

The `tempfile.mkstemp('.ending')` function allows us to create a temporary file with a certain file ending. The function returns a tuple, in which the first element is an OS-level handle for the open file and the second is the pathname of the file. Because we are concerned only with the path of the file, we disregard the first element.

After we have obtained the created path, we can save the model's weights to that path. Keras/TensorFlow provides many other methods of saving a model that you may want to use depending on the application, though. Using `model.save_weights()` only saves the model's weights, but not other attributes like a reloadable architecture. You can save the entire model such that it is entirely reloadable for inference or further training with `keras.models.save_weights(model, path)`. Set `include_optimizer` to `False` if the optimizer is not needed (the model is reloaded only for inference, not for further training).

The function can be defined using these components as follows (Listing 4-5).

Listing 4-5. Writing function to get the size to store a model

```
def get_size(model):

    # create file for weights
    _, weightsfile = tempfile.mkstemp(".h5")

    # save weights to file
    model.save_weights(weightsfile)

    # create file for zipped weights file
    _, zippedfile = tempfile.mkstemp(".zip")

    # zip weights file
    with zf.ZipFile(zippedfile, "w",
                    compression=zf.ZIP_DEFLATED) as f:
        f.write(weightsfile)

    # return size of model, in megabytes
    return str(os.path.getsize(zippedfile)/float(2**20))+' MB'
```

To obtain the storage required for a model, we simply pass the model object as a parameter into the `get_size` function.

We can compare the megabytes of storage required for the zipped unpruned model to the storage required for the zipped pruned model. Because of fixed storage requirements and variations in how different model architectures and other attributes can be stored, the outcome of pruning on storage requirements can vary.

Latency

Although latency can be calculated in many ways with many adaptations for particular applications, in this case the latency of a network simply refers to the average quantity of time the network takes to predict on a previously unseen sample (Listing 4-6).

Listing 4-6. Writing function to get the latency of a model

```
import time
def get_latency(model):
    start = time.time()
    res = model.predict(x_test)
    end = time.time()
    return (end-start)/(len(x_test))
```

Although in some cases it may not matter, it's good practice to make conscious decisions about the separation of training and deployment. In this case, latency is a metric that aims to understand how quickly the model can inference in a deployed environment, meaning it will inference on data it has not seen before. These decisions allow for mental clarity.

Parameter Metrics

The number of parameters isn't really an ends-oriented metric, meaning that the number of parameters cannot be used to precisely indicate the practical costs of storing or running a model. However, it's useful in that it is a direct measurement of pruning's effects on the number of parameters in a model. Note that while storage and latency are applicable across all compression methods, counting the number of pruned parameters compared to the number of parameters in the original model applies only to pruning.

You can obtain a list of a model's weights with `model.get_weights()`. For Sequential models, indexing the *i*th layer corresponds to the weights in the *i*th layer. Calling `np.count_nonzero()` on a layer's weights returns the number of nonzero parameters in that layer. It's important to count the number of nonzero parameters rather than the number of parameters; recall that in practice a pruned weight is simply set to 0.

We can thus find the total number of parameters in a model using list comprehension: `sum([np.count_nonzero(l) for l in orig_model.get_weights()])`. Using the parameter counts for the original and the pruned model, we can obtain a pruned-to-original weights ratio, indicating what fraction of original weights were retained in pruning, as well as the compression ratio, indicating what fraction of the original weights were pruned away (Listing 4-7).

Listing 4-7. Writing function to get parameter metrics

```
from numpy import count_nonzero as nz
def get_param_metrics(orig_model, pruned_model):

    orig_model_weights = orig_model.get_weights()
    om_params = sum([np.nz(l) for l in orig_model_weights])

    p_model_weights = pruned_model.get_weights()
    p_params = sum([np.nz(l).size for l in p_model_weights])

    return {'Original Model Parameter Count:': om_params,
            'Pruned Model Parameter Count': p_params,
            'Pruned to Original Weights Ratio': p_params/om_params,
            'Compression Ratio': 1 - p_params/om_params}
```

This function offers a simple and quick way to compare the number of parameters before and after pruning.

Pruning an Entire Model

Let's begin by importing the TensorFlow Model Optimization library as its commonly used abbreviation, `tfmot` (Listing 4-8).

Listing 4-8. Importing TensorFlow Model Optimization

```
import tensorflow_model_optimization as tfmot
```

To begin, we need to provide several parameters for pruning:

- *Initial sparsity*: The initial sparsity to begin with. For instance, an initial sparsity 0.50 indicates that the network begins with 50% of its parameters pruned.

- *Final sparsity*: The final sparsity to be reached after pruning is completed. For instance, a final sparsity of 0.95 indicates that when pruning is completed, 95% of the network is pruned.

- *Begin step*: The step to begin with. This is usually 0 to prune with the entire data.

- *End step*: The number of steps to train the data on.

- *Frequency*: The frequency with which to perform pruning (i.e., the network is pruned every [frequency] steps).

Here, a step indicates a batch, since the network generally performs an update after every batch. Given that the beginning step is 0, the end step indicates the total number of batches the network should run through during training. We can calculate it as

$$end\ step = ceil\left(\frac{training\ data\ length}{batch\ size}\right) \cdot epochs\ .$$ (Note that the default batch size in Keras is 32.)

These parameters will be passed into a pruning schedule (Listing 4-9). In this case, we use polynomial decay, in which weights are successively pruned in polynomial fashion such that the percent of weights pruned increases from the initial sparsity to the final sparsity. The update frequency should be small enough such that each increase in sparsity during pruning is not too large, but large enough such that the network has time to adapt to the pruning operation. In this case, we begin with 50% sparsity and work toward pruning away 95% of the parameters in the network.

Listing 4-9. Creating a polynomial decay schedule for pruning

```
from tfmot.sparsity.keras import PolynomialDecay as PD
schedule = PD(initial_sparsity=0.50,
              final_sparsity=0.95,
              begin_step=0,
              end_step=end_step,
              frequency=128)
```

TensorFlow Model Optimization also offers the `ConstantSparsity` schedule (`tfmot.sparsity.keras.ConstantSparsity`), which maintains constant sparsity throughout training. Rather than slowly increasing the percent of pruned parameters, constant sparsity keeps the same sparsity throughout training. This may be more optimal for simpler tasks, although polynomial decay is generally preferred, since it allows the network to adapt to the pruned parameters.

This schedule can be passed into a parameter dictionary (Listing 4-10). This parameter dictionary is unpacked and used, along with the model to be pruned, as a parameter in the `sparsity.prune_low_magnitude` function, which automatically prunes weights of low magnitude.

Listing 4-10. Creating a pruned model with pruning parameters. If you are unfamiliar, the ** kwargs syntax in Python passes the dictionary keys and values as parameter inputs to a function

```
pruning_params = {
    'pruning_schedule': schedule
}
pruned_model = tfmot.sparsity.keras.prune_low_magnitude(model,
**pruning_params)
```

Recall that, in pruning, the model should be substantially trained on the data before pruning begins. However, we are basing our pruned model on the original unpruned model, which we have already pretrained. The weights are transferred. If you do not perform pretraining, you will likely see worse results from pruning.

This model can be treated like a standard Keras model. Before training, it needs to be compiled, like any Keras model (Listing 4-11).

Listing 4-11. Compiling a pruned model

```
pruned_model.compile(loss='categorical_crossentropy',
                     optimizer='adam',
                     metrics=['accuracy'])
```

To perform the pruning step, we need to use the `UpdatePruningStep()` callback. This callback can be used in fitting (Listing 4-12).

Listing 4-12. Fitting a pruned model with the Update Pruning Step callback

```
update_pruning = tfmot.sparsity.keras.UpdatePruningStep()
pruned_model.fit(x_train, y_train,
                 epochs=15,
                 callbacks=[update_pruning])
```

During the process of pruning, TensorFlow Model Optimization automatically adds parameters to assist in pruning – each parameter is masked. If you count the number of parameters for a model at this stage, you'll notice it is significantly more than the original number of parameters.

To reap the fruits of pruning, use `tfmot.keras.sparsity.strip_pruning` to remove artifacts of the pruning training process: `pruned_model = tfmot.keras.sparsity. strip_pruning(pruned_model)`. This is necessary, along with a standard compression algorithm, to materialize the compression benefits.

After pruning is completed, it's best to fine-tune the model by recompiling and fitting it on the data again (Listing 4-13).

Listing 4-13. Fine-tuning after a model has been pruned

```
pruned_model.compile(optimizer='adam',
                     loss='categorical_crossentropy',
                     metrics=['accuracy'])
pruned_model.fit(x_train, y_train, epochs=10)
```

After fine-tuning, you can evaluate the performance of the `pruned_model` to understand the decrease in performance and improvement in compression and cost.

If you want to save the model, call `pruned_model.save(filepath)`. When reloading, make sure you reload the model under the `tfmot.sparsity.keras.prune_scope` scope, which allows for deserialization of the saved model (Listing 4-14).

Listing 4-14. Fine-tuning after a model has been pruned

```
with tfmot.sparsity.keras.prune_scope():
    pruned_model = keras.models.load_model(filepath)
```

If you only save weights (`model.save_weights()`), read a model pruned via the model checkpoint callback, or use Saved Model (`tf.saved_model.save(model, filepath)`), deserialization under the pruning scope is not necessary.

Pruning Individual Layers

Recall that when pruning the entire model, we called tfmot.keras.sparsity.`prune_low_magnitude()` on the entire model. One approach to prune individual layers is to call tfmot.keras.sparsity.`prune_low_magnitude()` on individual layers as they are being compiled. This is compatible with layer objects in both the Functional and Sequential API.

In this example neural network, we prune all layers other than the first and last Dense layers after the Input layer and before the output layer (Listing 4-15). When choosing which layers to prune, avoid ambitiously pruning initial layers responsible for feature extraction and layers critical to the knowledge-building abilities of the model.

Listing 4-15. Pruning individual layers by adding wrappers around layers. Activations are left out for the purpose of brevity

```
from tfmot.sparsity.keras import prune_low_magnitude as plm

pruned_model = keras.Sequential()
pruned_model.add(L.Input((784,)))
pruned_model.add(L.Dense(2**9))
pruned_model.add(plm(L.Dense(2**8), **pruning_params))
pruned_model.add(plm(L.Dense(2**7), **pruning_params))
pruned_model.add(plm(L.Dense(2**6), **pruning_params))
pruned_model.add(plm(L.Dense(2**5)))
pruned_model.add(L.Dense(10, activation='softmax'))
```

The benefit of pruning layers independently is that you can use different pruning schedules for different layers, for instance, by less ambitiously pruning layers that have less parameters to begin with. The model can then be compiled and fitted with the UpdatePruningStep() callback as discussed earlier and fine-tuned afterward.

However, the disadvantage of this method of selecting layers to prune is that you can't do any pretraining before pruning, since layers are wrapped in the pruning wrapper from they are defined. This results in a worse outcome than if the model was pretrained on data before pruning. To select specific layers for pruning on a model that has already been trained, we need to *clone* the model using keras.models.clone_model(model) with a cloning function. The cloning function maps each layer to another layer; in this case, we can map layers that we want to prune with a pruned version of that layer (Figure 4-5).

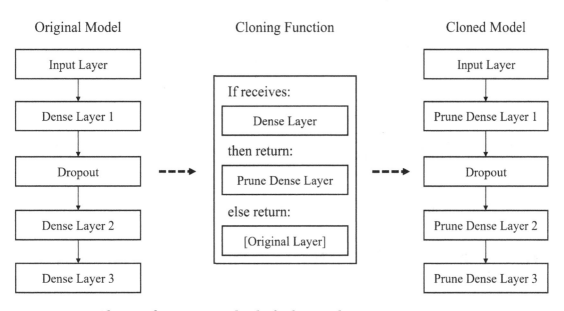

Figure 4-5. *Cloning function method of selecting layers to prune*

Let's construct a cloning function that either maps a layer to a pruned version of itself or returns the original layer if we do not desire to perform pruning on it (Listing 4-16). There are many ways you can select which layers to quantize; you can prune certain types of layers, layers by name, their position in the network, and so on. If a layer satisfies a condition for pruning, we return the layer wrapped in a pruning wrapper. Otherwise, we return the layer as is, untouched.

Listing 4-16. Defining a cloning function to map a layer to the desired state

```
def cloning_func(layer):

    # is it a Dense layer?
    if isinstance(layer, keras.layers.Dense):
        return plm(layer)

    # does it have a certain name?
    if layer.name == 'dense5':
        return plm(layer)

    # if does not meet any conditions for pruning
    return layer
```

Using this function, we can annotate the model by passing it as a cloning function when cloning the original model (Listing 4-17).

Listing 4-17. Using the cloning function with Keras' clone_model function

```
pruned_model = keras.models.clone_model(
    model,
    clone_function = cloning_func
)
```

Then, compile and fit (with the Update Pruning Step callback) as usual.

Pruning in Theoretical Deep Learning: The Lottery Ticket Hypothesis

Pruning is an especially important method not only for the purposes of model compression but also in advancing theoretical understandings of deep learning. Jonathan Frankle and Michael Carbin's 2019 paper "The Lottery Ticket Hypothesis: Finding Sparse, Trainable Neural Networks"[2] builds upon the empirical success of

---

[2] Jonathan Frankle and Michael Carbin, "The Lottery Ticket Hypothesis: Finding Sparse, Trainable Networks," 2019. Link to paper: `https://arxiv.org/pdf/1803.03635.pdf`. Link to code: `https://github.com/google-research/lottery-ticket-hypothesis`.

pruning to formulate the Lottery Ticket Hypothesis, a theoretical hypothesis that reframes how we look at neural network knowledge representation and learning.

Pruning has demonstrated that the number of parameters in neural networks can be decreased by upward of 90% with little damage to performance metrics. However, a prerequisite of pruning is that pruning must be performed on a large model; a small, trained network the same size as a pruned network still will not perform as well as the pruned network. A key component of pruning, it has been observed, is the element of *reduction*; the knowledge must first be learned by a large model, which is then iteratively reduced into fewer parameters. One cannot begin with an architecture mimicking that of a pruned model and expect to yield results comparable to the pruned model. These findings are the empirical motivation for the Lottery Ticket Hypothesis.

The Lottery Ticket Hypothesis states that initialized networks contain subnetworks that, when trained in isolation, reach performance comparable to the original network with a similar quantity of training. These winning subnetworks are referred to as "winning tickets." It is formally presented in Frankle and Carbin's paper as follows:

> *A randomly-initialized, dense neural network contains a subnetwork that is initialized such that—when trained in isolation—it can match the test accuracy of the original network after training for at most the same number of iterations.*

The primary contribution of the Lottery Ticket Hypothesis is in explaining the role of weight initialization in neural network development: weights that begin with convenient initialization values are "picked" up by optimizers and "developed" into playing meaningful roles in the final, trained network. As the optimizer determines how to update certain weights, certain subnetworks within the neural network are delegated to carry most of the information flow simply because their initialized weights were the right values to spark growth. On the other hand, weights that begin with poor initialization values are dimmed down as inconveniences and superfluous weights; these are the "losing tickets" that are pruned away in pruning. Pruning reveals the architecture containing the "winning tickets." Neural networks are running giant lotteries; the "winners" are amplified and the "losers" are attenuated.

You can think of a neural network from this perspective as a massive delivery package with a tiny valuable product inside and lots of stuffing. The vast majority of value is in a small minority of the actual package, but you need the package in the first place to find the product inside. Once you have the product, though, there's no need for the box anymore. Correspondingly, given that the initialization values are key to a network's success, you can retrain the pruned model architecture with the same corresponding initialization values and obtain similar performance to that of the original network.

This hypothesis reframes how we look at the training process of neural networks. The conventional perspective of machine learning models has always been that models begin with a set of "bad" parameters (the "initial guess") that are iteratively improved by finding updates that make the largest decrease to the loss function. With the largeness and even possible "overparameterization" of modern neural networks, though, the Lottery Ticket Hypothesis hints at a new logic of understanding training: learning is primarily a process not only of *improving* but also of *searching*. Promising subnetworks are developed via an alternating pattern of searching for promising subnetworks and improving promising subnetworks to become more promising. This new perspective toward understanding parameter updates in the large context of modern deep learning may fuel further innovation in theoretical understandings and in practical developments. For instance, we understand now that the initialization of weights plays a key role in the success of a subnetwork, which may direct further research toward understanding how weight initialization operates with respect to trained subnetwork performance.

The Lottery Ticket Hypothesis explains many observed phenomena in deep learning beyond the success and dynamics of pruning:

- It has been observed often that increasing the parametrization of neural networks leads to increased performance. The Lottery Ticket Hypothesis tells us that overparameterization is not necessarily inherently tied to greater predictive power, but that networks with larger quantities of parameters are able to run larger lotteries that yield better and more winning tickets. If the Lottery Ticket Hypothesis is true, it may provide a North Star for how to improve the quality of winning tickets rather than brute-force increasing the size of the lottery operation.

- It has been observed that initializing all weights to 0 performs much worse than other initialization methods that randomize weights. The Lottery Ticket Hypothesis tells us that networks rely upon a diversity of initial randomized weights in order to select for certain winning tickets. If all the weights are 0, the network cannot differentiate promising subnetworks from the start.

Because pruning strips away "losing tickets," Frankle and Carbin propose a pruning-based method to identify winning tickets:

1. Randomly initialize a neural network.

2. Train the neural network until convergence.

3. Prune away p% of the parameters in the trained neural network.

4. Reset the unpruned parameters to their original initialization values.

The Lottery Ticket Hypothesis and undoubtedly further theoretical advances in our understanding of neural networks guided by empirically observed phenomena in model compression will continue to serve as stepping stones to accelerating the improvement of our model-building methods.

Quantization

While pruning decreases the number of parameters, quantization decreases the precision of each one. Because each parameter that has been quantized is less precise, the model as a whole requires less storage and has decreased latency. The process of implementing quantization with TensorFlow Model Optimization is very similar to that of implementing pruning.

Quantization Theory and Intuition

Traditionally, neural networks use 32 bits to represent parameters; while this is fine for training in modern deep learning environments that have the computing power to use such a precision, it is not feasible in applications that require lower storage and faster predictions. In quantization, parameters are reduced from their 32-bit representations to 8-bit integer representations, leading to a fourfold decrease in memory requirements.

In mathematics, quantization is the mapping of a continuous set of values to a smaller set of discrete values (Figure 4-6). In deep learning, quantization refers to a wide set of methods that can be used to reduce the precision of a parameter via a similar method. Generally, this is performed by separating values into information buckets. In binary quantization, values are quantized into two buckets; in ternary quantization, values are quantized into three buckets. However, binary and ternary quantization may be too extreme, which is why most deployed models employ a multiple-bit-to-multiple-bit quantization approach. How these bins are placed, how large each bin is, and other parameters to perform this mapping are dependent on which quantization strategy is being used.

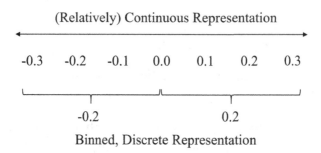

Figure 4-6. *Continuous vs. binned, discrete representation*

(You could view magnitude-based pruning as a selective form of quantization, in which weights smaller in magnitude than a certain threshold are "quantized to 0" and other weights are binned to themselves.)

Recall the living space analogy for pruning. Instead of outright removing certain items from your living space, imagine reducing the cost to keep each one by a little bit. You decide to downgrade your television subscription to a lower tier, decrease the electricity consumption of your light, order takeout once a week instead of twice or thrice a week, and other amends that "round" the cost of your experience down.

Post-processing quantization is a quantization procedure performed on the model after it is trained. While this method achieves a good compression rate with the advantage of decreased latency, the errors in the small approximations performed in each of the weights via quantization accumulate and lead to a significant decrease in performance.

Like the iterative approach to pruning, quantization is generally not performed on the entire network at once – this is too jarring a change, just as pruning away 95% of a network's parameters all at once is not conducive to recovery. Rather, after a model is pretrained – ideally, pretraining develops meaningful and robust representations that can be used to help the model recover from the compression – the model undergoes Quantization Aware Training, or QAT (Figure 4-7).

Throughout Quantization Aware Training, the model itself remains unquantized, representing all of its parameters with the standard 32 bits. However, a quantization error is introduced for consideration: in the network's feed-forward stage, the output of the network is the same as if the network had been quantized. That is, before any prediction, the network undergoes "simulated quantization" – for the purposes of prediction, its parameters are quantized. This simulated quantization output is used to update the model parameters, which are still unquantized. Thus, while the model itself remains unquantized throughout Quantization Aware Training, it learns to develop parameters that will succeed *when the model becomes quantized*. The model is left unquantized because it is significantly easier to update model parameters with more precise parameters.

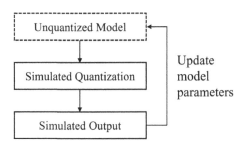

Figure 4-7. *Quantization Aware Training*

After Quantization Aware Training, the model is formally quantized – its parameters are binned and it uses the 8-bit integer representation (or some other representation, depending on the implementation). Because of Quantization Aware Training's preparation, the model should have developed parameters that are robust and successful when quantized (Figure 4-8).

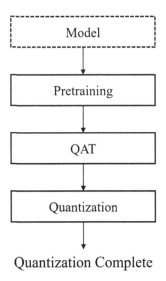

Figure 4-8. *Quantization process*

With quantization, a model's storage requirements and latency can be dramatically decreased with little effect on performance.

Quantization Implementation

Like in pruning, you can quantize an entire model or quantize layers independently.

Quantizing an Entire Model

Quantization requires pretraining for optimal performance. Let's begin by fitting a large base model on MNIST data for 15 epochs (Listing 4-18).

Listing 4-18. Base model for MNIST data; this will be used for applying quantization

```
import keras
import keras.layers as L

model = keras.Sequential()
model.add(L.Input((784,)))
for i in list(range(5,10))[::-1]:
    model.add(L.Dense(2**i, activation='relu'))
    model.add(L.Dense(2**i, activation='relu'))
model.add(L.Dense(10, activation='softmax'))
model.compile(optimizer='adam',
              loss='categorical_crossentropy',
              metrics=['accuracy'])
model.fit(x_train, y_train, epochs=15)
```

This particular model obtains a training loss of 0.0387 and a training accuracy of 0.9918. In evaluation, it scores a loss of 0.1513 and an accuracy of 0.9720. This sort of disparity between training and testing performance indicates that some sort of compression method would be apt to apply here.

To perform Quantization Aware Training on an entire model, import the quantize_
model function from tfmot.quantization.keras and apply it to the model (Listing 4-19);
this performs a "quantization annotation" on each layer that allows for Quantization Aware
Training. Because this removes the optimizer from the model, we'll need to recompile it.

Listing 4-19. Setting up Quantization Aware Training

```
from tfmot.quantization.keras import quantize_model
qat_model = quantize_model(model)
qat_model.compile(optimizer='adam',
                  loss='categorical_crossentropy',
                  metrics=['accuracy'])
```

Call quantized_model.evaluate(x_test, y_test), and you'll notice that the model
performs poorly. Like adapting to a new living space, we will need to perform some
additional training on the quantized model. When performing this additional fine-
tuning, make sure you have a high batch size and train for a small number of epochs
(Listing 4-20). In low-precision training, small batch sizes cause aggressive weight
updates that blow up the loss without recovery. A few epochs of large batch training
should be enough to orient the model toward good performance.

Listing 4-20. Performing Quantization Aware Training

```
qat_model.fit(x_train, y_train,
              batch_size=512,
              epochs=3)
```

Now, the model is *quantization aware*, meaning that it possesses the necessary
facilities for quantization, but it's not technically quantized. To reap the benefits
of quantization, we will need to convert the model into a TFLite model, which is
TensorFlow's solution for lightweight applications (Listing 4-21).

Listing 4-21. Converting to TFLite model to actually quantize model

```
converter = tf.lite.TFLiteConverter.from_keras_model(
    qat_model)
converter.optimizations = [tf.lite.Optimize.DEFAULT]
quantized_tflite_model = converter.convert()
```

We can then save and zip our TFLite model to see storage benefits (Listing 4-22).

Listing 4-22. Realizing storage benefits from TFLite model

```
# store TFLite model
with open('model.tflite', 'wb') as f:
    f.write(quantized_tflite_model)

# zip the file the model is stored in
_, zippedfile = tempfile.mkstemp(".zip")
with zf.ZipFile(zippedfile, "w",
                compression=zf.ZIP_DEFLATED) as f:
    f.write('model.tflite')

# output size of model
str(os.path.getsize(zippedfile) / float(2 ** 20)) + ' MB'
```

Like the pruned model, you can store the model to a file path via a variety of model saving and weight saving methods. If you load the weights by saving the entire model directly, make sure to reload the model under the scope tfmot.quantization.keras.quantize_scope.

Quantizing Individual Layers

Like in pruning, quantizing individual layers gives the advantage of specificity and hence smaller performance degradation, with the cost of a likely smaller compression than a fully quantized model.

When selecting which layers can be quantized, you can use tfmot.quantization.keras.quantize_annotate_layer, which you can wrap around a layer as you're using it either in the Sequential or Functional API, much like prune_low_magnitude(). When quantizing individual layers, try to quantize later layers rather than the initial layers.

If you are deploying quantized models, keep in mind that some backends may support only fully quantized models. In this case, you would want to quantize the entire model rather than choosing certain layers to quantize (Listing 4-23).

Listing 4-23. Quantizing individual layers by wrapping quantization annotations to individual layers while defining them

```
from tfmot.quantization.keras import quantize_annotate_layer as qal

annotated_model = keras.Sequential()
annotated_model.add(L.Input((784,)))
annotated_model.add(qal(L.Dense(2**9)))
annotated_model.add(L.Activation('relu'))
annotated_model.add(qal(L.Dense(2**8)))
annotated_model.add(L.Activation('relu'))
annotated_model.add(qal(L.Dense(2**7)))
annotated_model.add(L.Activation('relu'))
annotated_model.add(L.Dense(2**6, activation='relu'))
annotated_model.add(L.Dense(2**5, activation='relu'))
annotated_model.add(L.Dense(10, activation='softmax'))
```

Note that, at this point, the layers you applied the `quantize_annotate_layer` are only annotated. To convert them into actually quantized layers, we need to use `quantize_apply` (Listing 4-24).

Listing 4-24. Applying quantization to the annotated layers

```
from tfmot.quantization.keras import quantize_apply
quantized_model = quantize_apply(annotated_model)
```

The `quantize_apply` function was not needed when quantizing the entire model using `quantize_model` because the `quantize_model` function acts as a "shortcut" that annotated and applies quantization automatically for general cases in which the "default" parameters can be applied (i.e., there is no need for customization by quantizing specific layers).

The model can then be compiled and fitted using the same training principles as discussed prior – low number of epochs, high batch size.

Like in pruning, the preferred method of selecting layers to quantize is to define a cloning function and use `keras.models.clone_model(model)` (Listing 4-25).

Listing 4-25. Defining a quantization annotation cloning function

```
def cloning_func(layer):

    # is it a Dense layer?
    if isinstance(layer, keras.layers.Dense):
        return qal(layer)

    # does it have a certain name?
    if layer.name == 'dense5':
        return qal(layer)

    # if does not meet any conditions for quantization
    return layer
```

Using this function, we can annotate the model by passing it as a cloning function when cloning the original model (Listing 4-26).

Listing 4-26. Applying the cloning function to a (pretrained) base model

```
annotated_model = keras.models.clone_model(
    model,
    clone_function = cloning_func
)
```

Then, apply the `quantize_apply` function to the annotated model and compile and fit like normal.

Weight Clustering

Weight clustering is a certainly less popular but still incredibly valuable and simple model compression method (Figure 4-9).

Weight Clustering Theory and Intuition

Weight clustering is a combination of pruning and quantization in character – it reduces the number of *unique* weights by slightly adjusting each weight value. Given a user-specified quantity of clusters n, the weight clustering algorithm assigns each weight value a cluster and sets the weight value to the centroid of that weight value (Figure 4-9).

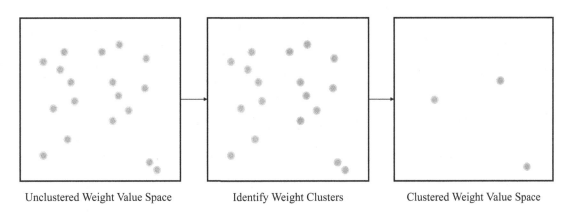

Unclustered Weight Value Space   Identify Weight Clusters   Clustered Weight Value Space

Figure 4-9. *Weight clustering*

Weights that are part of one cluster all share the same value, thus allowing for more efficient means of storage. Similarly to quantization, the decrease in storage requirement is a matter of precision; each parameter's precise value can be replaced by the index of the associated centroid value. These precise values can be stored once in an indexable list of centroid values (Figure 4-10). (Note that even if this method of centroid indexing is not used, compression algorithms will be able to take advantage of repeated values.)

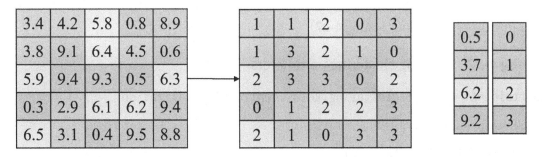

Unclustered Weight Matrix   Clustered Weight Matrix   Indexing Values

Figure 4-10. *Weight clustering via indexing*

The key parameter in weight clustering is determining the number of clusters. Like the percent of parameters to prune, this is a trade-off between performance and compression. If the number of clusters is very high, the change each parameter makes from its original value to the value of the centroid it was assigned is very small, allowing for more precision and easier recovery from the compression operation. However, it diminishes the compression result by increasing storage requirements both for storing the centroid values and potentially the indexes themselves. On the other hand, if the number of clusters is too small, the performance of the model may be so impaired that it cannot recover – it may simply not be possible for a model to function reasonably with a certain set of fixed parameters.

Weight Clustering Implementation

Like with pruning and quantization, you can either cluster the weights of an entire model or of individual layers.

Weight Clustering on an Entire Model

Like pruning and quantization, weight clustering requires a pretrained `model`. In order to perform weight clustering on the model, we first need to provide clustering parameters. There are two key parameters to provide: the number of clusters and the method of centroid initialization. Although in this example the chosen method of initialization is density-based sampling, you can also use `CentroidInit.LINEAR`, in which cluster centroids are evenly spaced between minimum and maximum values; `CentroidInit.RANDOM`, in which centroids are randomly sampled from a uniform distribution between the minimum and maximum values; and `CentroidInit.KMEANS_PLUS_PLUS`, which uses the K-means++ algorithm (Listing 4-27).

Listing 4-27. Defining clustering parameters

```
CentroidInit = tfmot.clustering.keras.CentroidInitialization

clustering_params = {
    'number_of_clusters': 30,
    'cluster_centroids_init': CentroidInit.DENSITY_BASED
}
```

To perform clustering on an entire model, use the `cluster_weights()` function within `tfmot.clustering.keras` with the specified parameters (Listing 4-28).

Listing 4-28. Creating a weight-clustered model with the specified clustering parameters

```
from tfmot.clustering.keras import cluster_weights
clustered_model = cluster_weights(model, **clustering_params)
```

The weight-clustered model can then be compiled and fitted on the original data for fine-tuning.

In order to realize the compression benefits of clustering, use `strip_clustering()` to clear the model of any artifacts from weight clustering (Listing 4-29).

Listing 4-29. Stripping clustering artifacts to realize compression benefits after fitting

```
from tfmot.clustering.keras import strip_clustering
final_model = strip_clustering(clustered_model)
```

After this, convert the code into a TFLite model and evaluate the size of the zipped TFLite model to see the decrease in storage size. You can also evaluate the latency of the model by using the function we defined prior in the pruning section, but make sure to re-attach an optimizer by compiling.

Like a pruned and quantized model, you can store the weight-clustered model to a file path via a variety of model saving and weight saving methods. If you load the weights by saving the entire model directly, make sure to reload the model under the scope `tfmot.clustering.keras.cluster_scope`.

Weight Clustering on Individual Layers

Weight clustering on individual layers follows the same syntax as pruning and quantization on individual layers, but use `tfmot.clustering.keras.cluster_weights` instead of `tfmot.quantization.keras.quantize_apply` or `tfmot.sparsity.prune_low_magnitude`. Like these other compression methods, you can either apply weight clustering to each layer as it is being constructed in the architecture or as a cloning function when cloning an existing model. The latter procedure of applying a compression method to individual layers is preferred because it allows for convenient pretraining and fine-tuning.

Collaborative Optimization

Generally, you can obtain good results using compression methods individually. When these compression methods are combined, though, you can achieve increased compression with better performance: the fundamental idea behind *collaborative optimization* is that compression methods can be chained together such that each acts to compress the model in its own unique method to achieve a more successful net compression than if just one (proportionally scaled) compression method had been applied (Figure 4-11). Practical deployment of deep learning almost always employs collaborative optimization rather than one compression method in isolation.

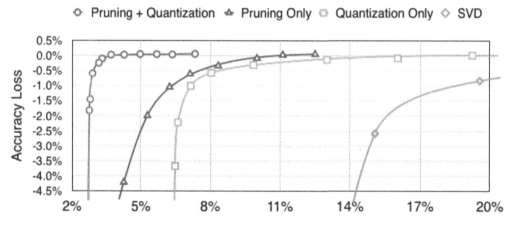

Figure 4-11. *Relationship between model size ratio after compression and the loss in accuracy by compression method. For a certain accuracy loss, pruning + quantization is able to achieve a much smaller model size ratio after compression than pruning only or quantization only. SVD is another model compression technique that has not seen as much success as pruning and quantization*

Given the three compression methods that have been discussed, there are three two-method combinations:

- Quantization and weight clustering, or clustering preserving quantization

- Quantization and pruning, or sparsity preserving quantization

- Pruning and weight clustering, or sparsity preserving clustering

The naming of these methods is significant, because it implies a certain order in which these operations are applied. For instance, if we were to apply weight clustering and quantization, it would be optimal to apply weight clustering first and then quantization rather than vice versa. When using collaborative optimization, there is generally an "order of operations":

```
pruning, weight clustering, quantization
```

These are ordered such that each compression method interferes as little as possible with other compression methods. Pruning and weight clustering, for instance, require relatively high-precision information and would be severely disrupted if quantization were performed first. Pruning relies upon the existence of a wide, diverse array of parameters to rank and choose from; if weight clustering were performed before pruning, it would significantly decrease the diversity of values and therefore disrupt the efficacy of pruning (Figure 4-12).

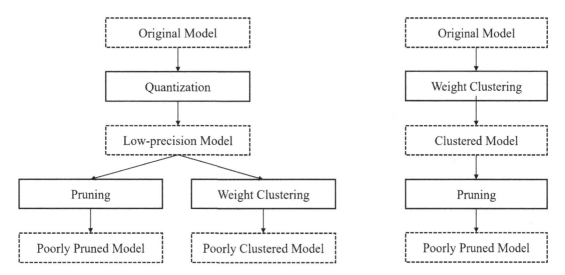

Figure 4-12. *The effect of the order in which model compression methods are applied on the performance of the collaboratively optimized model. Performing quantization before pruning or weight clustering and weight clustering before pruning undermines the effect of the second compression method and therefore is an inefficient process*

However, when applying collaborative optimization, you cannot simply apply one method after another. Even given our "order of operations" to optimize the performance of chained methods, in practice adding an additional compression method severely dampens the effect of the previous one (Figure 4-13). For instance, consider weight clustering and pruning – pruning sets pruned parameters to zero, but weight clustering sets parameters to whatever their centroid value is. Thus, if weight clustering were performed after pruning, many of the pruned parameters would be "unpruned" because they were set to a nonzero centroid value.

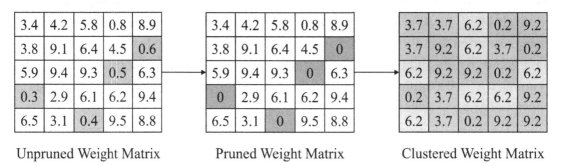

Unpruned Weight Matrix   Pruned Weight Matrix   Clustered Weight Matrix

Figure 4-13. *The undoing effect of performing weight clustering after pruning without using sparsity preserving clustering. Even though the difference in this case is small, it can be compounded significantly across each of the weight matrices for tremendous damage to the effects of pruning*

Thus, specialized versions of quantization and clustering are needed to perform their respective compression methods while maintaining the compression effect of the previous method (Figure 4-14).

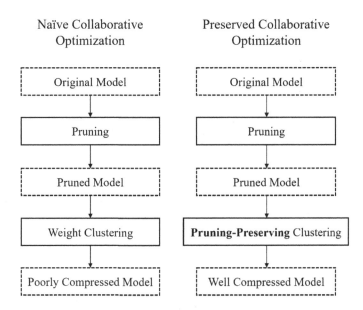

Figure 4-14. *The importance of model compression preservation in collaborative optimization*

Sparsity Preserving Quantization

In sparsity preserving quantization, pruning is followed by quantization (Figure 4-15).

Using code and methods discussed earlier in the pruning section, obtain a pruned_model. You can use the measurement metrics defined prior to verify that the pruning procedure was successful. Use the strip_pruning function (tfmot.sparsity.keras. strip_pruning) to remove artifacts from the pruning procedure; this is necessary in order to perform quantization.

Recall that to induce Quantization Aware Training for a model, you used the quantize_model() function and then compiled and fitted the model. Performing pruning-preserving Quantization Aware Training, however, requires an additional step. The quantize_annotate_model() function is used not to actually quantize the model, but to provide annotations indicating that the entire model should be quantized. quantize_annotate_model() is used for more specific customizations of the quantization procedure, whereas quantize_model() can be thought of as the "default" quantization method. (You may similarly recall that quantize_annotate_layer() was used for another specific customization – layer-specific quantization.)

After the entire model has been annotated, we use the quantize_apply() function to actually quantize the annotated model. In this function, we can specify the preservation of another compression method – in this case, pruning. This is specified by passing a tfmot.experimental.combine object, which indicates a compression method to be preserved when "combining" or "collaborating." The pruning-preserving Quantization Aware Training model can then be compiled and fitted as usual.

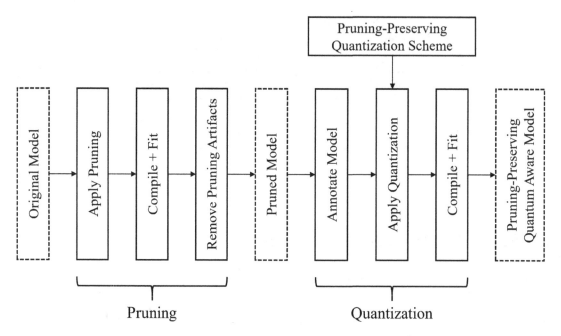

Figure 4-15. *Collaborative optimization with sparsity preserving quantization*

The complete code is as follows (Listing 4-30).

Listing 4-30. Performing sparsity preserving quantization after pruning

```
# removing pruning artifacts for quantization
from tfmot.pruning.keras import strip_pruning
pruned_model = strip_pruning(pruned_model)

# annotate entire model
from tfmot.quantization.keras import quantize_annotate_model
annot_quant_model = quantize_annotate_model(pruned_model)
```

```
# specify combining method (pruning)
from tfmot.experimental.combine import
Default8BitClusterPreserveQuantizeScheme as preserve_pruning

# apply quantization to annotated model
from tfmot.quantization.keras import quantize_apply
pqat_model = quantize_apply(annot_quant_model,
                           preserve_pruning())

# compile and fit
pqat_model.compile(...)
pqat_model.fit(...)
```

Cluster Preserving Quantization

In cluster preserving quantization, weight clustering is followed by quantization (Figure 4-16).

Using code and methods discussed earlier in the "Weight Clustering" section, obtain a clustered_model. From here, the process is almost the same as sparsity preserving quantization: after stripping clustering artifacts from the clustered_model, annotate the model and use quantize_apply to quantize the annotated layers. When specifying which compression method to preserve in quantize_apply, use Default8BitClusterPreserveQuantizeScheme rather than Default8BitPrunePreserveQuantizeScheme.

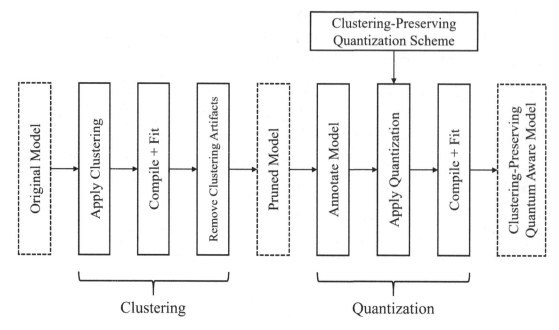

Figure 4-16. *Collaborative optimization with cluster preserving quantization*

Sparsity Preserving Clustering

In sparsity preserving clustering, pruning is followed by weight clustering (Figure 4-17).

Sparsity preserving clustering follows a slightly different process than cluster preserving quantization and sparsity preserving quantization.

Using code and methods discussed earlier in the pruning section, obtain a `pruned_model`. Strip away pruning artifacts with `strip_pruning`.

We need to import the `cluster_weights` function to perform weight clustering; prior, we imported it from `tfmot.clustering.keras.cluster_weights`. However, to use sparsity preserving clustering, we need to import the function from a different place: from tensorflow_model_optimization.python.core.clustering.keras.experimental.cluster import cluster_weights.

Now, we can provide weight clustering parameters, as before, with one additional "`preserve_sparsity`" argument (Listing 4-31).

Listing 4-31. Defining clustering parameters with sparsity preservation marked

```
# specify centroid initialization style
from tfmot.clustering.keras import CentroidInitialization
CentroidInit = CentroidInitialization.DENSITY_BASED
```

```
# put clustering parameters into dictionary
clustering_params = {'number_of_clusters': 8,
                     'cluster_centroids_init': CentroidInit,
                     'preserve_sparsity': True}
```

Then, apply the cluster_weights function to the stripped pruned model with the clustering parameters, and compile and fit (Listing 4-32).

Listing 4-32. Performing sparsity preserving clustering after pruning

```
# create sparsity preserving clustering model
spc = cluster_weights(pruned_model, **clustering_params)

# compile and fit
spc.compile(...)
spc.fit(...)
```

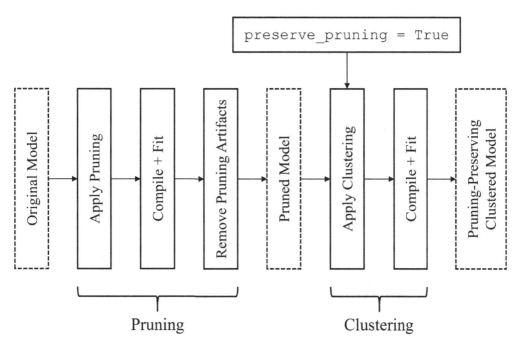

Figure 4-17. *Collaborative optimization with sparsity preserving clustering*

Case Studies

In these case studies, we will present research that has experimented with these compression methods and other variations on the presented methods to provide further concrete exploration into model compression.

Extreme Collaborative Optimization

The 2016 paper "Deep Compression: Compressing Deep Neural Networks with Pruning, Trained Quantization, and Huffman Coding"[3] by Song Han, Huizi Mao, and William J. Dally was an instrumental leap forward in collaborative optimization.

The paper proposed a three-stage compression pipeline: pruning, weight clustering and quantization (grouped together as one method), and Huffman coding (Figure 4-18). This compression pipeline progressively compresses large models like AlexNet and VGG-16 on the ImageNet dataset between 35 times and 49 times without incurring any loss in accuracy. Moreover, the latency is decreased by three to four times with three to seven times improved energy efficiency. By chaining compression methods in this order, the compression methods minimally interfered with one another, leading to surprisingly large compression:

1. Recall that pruning is best performed as an iterative process in which connections are pruned and the network is fine-tuned on those pruned connections. In this paper, pruning reduces the model size by 9 to 13 times with no decrease in accuracy.

2. Recall that weight clustering is performed by clustering weights that have similar values and setting the weights to their respective centroid values and that quantization is performed by training the model to adapt toward lower-precision weights. Weight clustering combined with quantization, after pruning, reduces the original model size by 27 to 31 times.

---

[3] Song Han, Huizi Mao, and William J. Dally, "Deep Compression: Compressing Deep Neural Networks with Pruning, Trained Quantization and Huffman Coding," 2016. Link to paper: https://arxiv.org/pdf/1510.00149.pdf. Link to code: https://github.com/songhan/Deep-Compression-AlexNet (Caffe implementation).

3. Huffman coding is a compression technique that was proposed
    by computer scientist David A. Huffman in 1952. It allows for
    lossless data compression that represents more common symbols
    with fewer bits. Huffman coding is different from the previously
    discussed model compression methods because it is a post-training
    compression scheme; that is, there is no model fine-tuning required
    for the scheme to work successfully. Huffman encoding allows for
    even further compression – the final model is compressed with a
    reduction 35 to 49 times its original size.

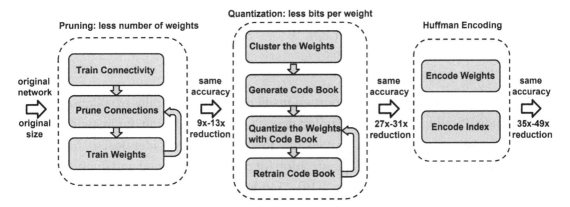

Figure 4-18. *Collaborative optimization between pruning, weight clustering,
quantization, and Huffman encoding*

This compression pipeline successfully compresses large architectures by several
dozen times with little effect on error, an incredible feat (Table 4-1).

Table 4-1. *Performance of this collaborative optimization compressed model on MNIST for LeNet and AlexNet and on ImageNet for VGG-16 model*

Network	Top 1 Error	Top 5 Error	Parameters	Compression Rate
LeNet-300-100	1.64%	–	1070 KB	40 times
Compressed	1.58%		27 KB	
LeNet-5	0.80%	–	1720 KB	39 times
Compressed	0.74%		44 KB	
AlexNet	42.78%	19.73%	240 MB	35 times
Compressed	42.78%	19.70%	6.9 MB	
VGG-16	31.50%	11.32%	552 MB	49 times
Compressed	31.17%	10.91%	11.3 MB	

Han, Mao, and Dally provide important insights into the dynamics of collaborative optimization. Pruning before quantization doesn't hurt quantization, for instance – the performance of a model both pruned and quantized is almost identical to that of a model that has only undergone quantization (of course, the pruned and quantized model has fewer parameters) (Figure 4-19). This demonstrates a key property of ideal collaborative optimization: strength is found in a diverse array of compression attacks. By chaining a diverse set of compression methods that each attack different representation redundancies, the model is stripped of inefficient representations from all "angles" and therefore results in higher compression while still maintaining the necessary essential facilities for good performance.

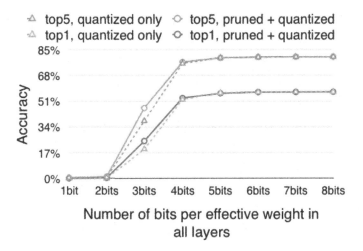

Figure 4-19. *Performance of models with various compression methods applied*

Rethinking Quantization for Deeper Compression

Recall that when performing quantization, Quantization Aware Training is used to orient the model toward learning weights that are robust to quantization. This is performed by simulating a quantized environment when the model is making predictions.

However, Quantization Aware Training raises a key problem: because quantization effectively "discretizes" or "bins" a functionally "continuous" weight value, the derivative with respect to the input is zero at almost everywhere, posing problems for the gradient update calculations. In order to work around this, in practice a Straight Through Estimator is used. As implied by its name, a Straight Through Estimator estimates the output gradients of the discretized layer as its input gradients without regard for the actual derivative of the actual discretized layer. A Straight Through Estimator works with relatively less aggressive quantization (i.e., the 8-bit integer quantization implemented earlier) but fails to provide sufficient estimation for more severe compressions (e.g., 4-bit integer).

To address this problem, Angela Fan and Pierre Stock, along with Benjamin Graham, Edouard Grave, Remi Gribonval, Herve Jegou, and Armand Joulin, propose quantization noise in their paper "Training with Quantization Noise for Extreme Model Compression,"[4] a novel approach toward orienting a compressed model to developing quantization robust weights.

Rather than simulating the quantization of the entire model, like in Quantization Aware Training, quantization noise instead simulated the quantization of *part of a model* – a randomly selected subset of weights are simulated quantized during each forward pass (Figure 4-20). This means that most of the weights are updated with cleaner gradients.

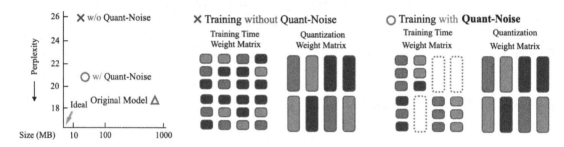

Figure 4-20. *Demonstration of training without quantization noise vs. with quantization noise*

Quantization noise significantly improves the performance of low-precision compression methods over Quantization Aware Training, both in the domains of language modeling and image classification (Table 4-2).

---

[4] Angela Fan, Pierre Stock, Benjamin Graham, Edouard Grave, Remi Gribonval, Herve Jegou, and Armand Joulin, "Training with Quantization Noise for Extreme Model Compression," 2021. Link to paper: `https://arxiv.org/pdf/2004.07320.pdf`. Link to code: `https://github.com/pytorch/fairseq/blob/master/examples/quant_noise/README.md` (PyTorch implementation).

Table 4-2. *Language modeling task: 16-layer transformer on WikiText-103. Image classification: EfficientNetB3 on ImageNet 1k. "Comp." refers to "Compression." "PPL" refers to perplexity, a metric for NLP tasks (lower is better). QAT refers to Quantization Aware Training; QN refers to quantization noise.*

Quantization Method	Language Modeling			Image Classification		
	Size	Comp.	PPL	Size	Comp.	Top 1
Uncompressed method	942	1x	18.3	46.7	1x	81.5
4-bit integer quantization	118	8x	39.4	5.8	8x	45.3
– Trained with QAT	118	8x	34.1	5.8	8x	59.4
– Trained with QN	118	8x	21.8	5.8	8x	67.8
8-bit integer quantization	236	4x	19.6	11.7	4x	80.7
– Trained with QAT	236	4x	21.0	11.7	4x	80.8
– Trained with QN	236	4x	18.7	11.7	4x	80.9

While fixed-point scalar quantization methods introduced in this chapter like int8 quantization reduce the precision of parameter values via "rounding," there exist other quantization methods. Fan and Stock also explore quantization noise on *product quantization*, a method in which a high-dimensional vector space is decomposed into several subspaces that are quantized separately. Like the rationale for Quantization Aware Training and Iterative Pruning, product quantization is best performed iteratively. This iterative product quantization (iPQ) method generally obtains higher compression rates than rounding to some bit-level precision (Table 4-3).

Table 4-3. *iPQ with quantization noise compared to the performance of an uncompressed model*

Quantization Method	Language Modeling			Image Classification		
	Size	Comp.	PPL	Size	Comp.	Top 1
Uncompressed method	942	1x	18.3	46.7	1x	81.5
iPQ	38	25x	25.2	3.3	14x	79.0
– Trained with QAT	38	25x	41.2	3.3	14x	55.7
– Trained with QN	38	25x	20.7	3.3	14x	80.0

Responsible Compression: What Do Compressed Models Forget?

When we talk about compression, the two key figures we consider are the model performance and the compression factor. These two figures are often balanced and used to determine the success of a model compression operation. We often see an increase in compression accompanied by a decrease in performance – but do you wonder what *types* of data inputs are being sacrificed by the compression procedure? What is lying beneath the generalized performance metrics?

In "What Do Compressed Deep Neural Networks Forget?",[5] Sara Hooker, along with Aaron Courville, Gregory Clark, Yann Dauphin, and Andrea Frome, investigates just this question: how do compression methods affect what knowledge a compressed model is forced to "forget" by the compression? Hooker et al.'s findings suggest that looking merely at standard performance metrics like test set accuracy may not be enough to reveal the impact of compression on the model's true generalization capabilities.

Pruning Identified Exemplars (PIEs) are defined inputs to a model in which there is a high level of disagreement between the predictions of pruned and unpruned models. Hooker et al. find that general metrics like test set accuracy hide important information regarding the effect of pruning on the model's generalization capabilities; model compression methods like pruning *do not uniformly affect the model's data to process instances across the distribution of the dataset*. Rather, a small subset of data is disproportionately impacted (Figure 4-21).

---

[5] Sara Hooker, Aaron Courville, Gregory Clark, Yann Dauphin, and Andrea Frome, "What Do Compressed Deep Neural Networks Forget?", 2020. Link to paper: https://arxiv.org/pdf/1911.05248.pdf. Link to code: https://github.com/google-research/google-research/tree/master/pruning_identified_exemplars.

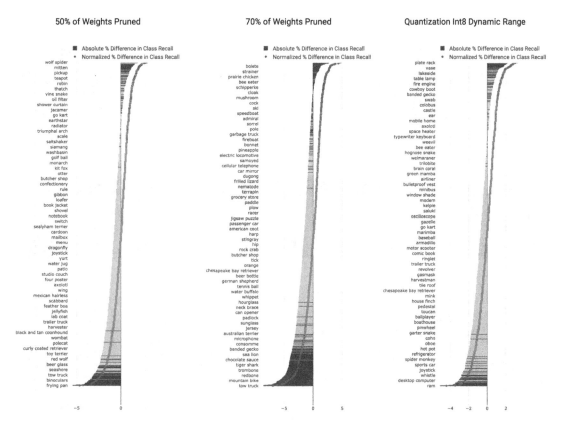

Figure 4-21. *Increase or decrease in a compressed model's recall for certain ImageNet classes. Colored bars indicate the classes in which the impact of compression is statistically significant. As a higher percent of weights are pruned, there are more classes that are statistically significantly affected by compression. Note that, interestingly, quantization suffers less from these generalization vulnerabilities than pruning does*

Data instances from the long tail of the dataset distribution – that is, less commonly represented or more complex data instances – are more often "sacrificed" in model compression. Hooker at al. asked human subjects to label components of Pruning Identified Exemplars and found that PIEs were more difficult both for humans and models to classify; PIEs are generally more complex, consisting of multiple objects, lower quality, or ambiguity (Figure 4-22). Pruning forces compressed models to sacrifice understanding of these particular instances, exposing vulnerabilities in compressed model generalization.

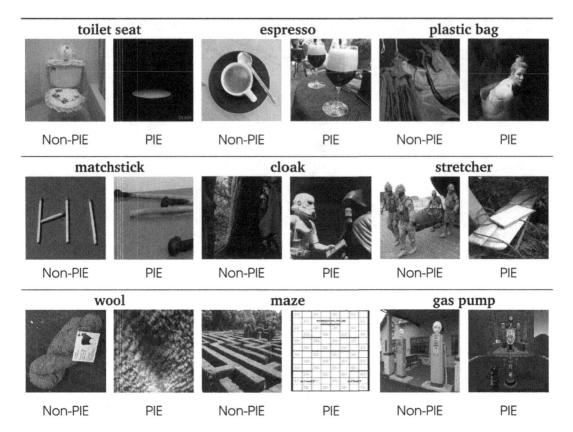

Figure 4-22. *Pruning Identified Exemplars are more difficult to classify and are less represented*

Compressed models, moreover, are more prone to small changes that humans would be robust to. The higher the compression, the less robust the model becomes to variations like changing brightness, contrast, blurring, zooming, and JPEG noise. This also increases the vulnerability of compressed models used in deployment to adversarial attacks, or attacks designed to subvert the model's output by making small, cumulative changes undetectable to humans (see Chapter 2, Case Study 1, on adversarial attacks exploiting properties of transfer learning).

In addition to posing concerns for model robustness and security, these findings raise questions for the role of model compression in increasing discussion on fairness. Given that model compression disproportionately affects the model's capacity to process less represented items in categories, model compression can amplify existing disparities in dataset representation.

Hooker et al.'s work reminds us that neural networks are complex entities that often may require more exploration and consideration than wide-reaching metrics may suggest and leaves important questions to be answered in future work on model compression.

Key Points

In this chapter, we discussed the intuition and implementation of three key model compression methods – pruning, quantization, and weight clustering – as well as collaborative optimization techniques by which these compression methods can be chained together – sparsity preserving quantization, cluster preserving quantization, and sparsity preserving clustering:

- The goal of model compression is to decrease the "cost" a model incurs while maintaining model performance as much as possible. The "cost" of a model encompasses many factors, including storage, latency, server-side computation and power cost, and privacy. Model compression is a core element both of practical deployment and key to advancing theoretical understandings of deep learning.

- In pruning, unimportant parameters or other more structured network elements are "removed" by being set to 0. This allows for much more efficient storage of the network. Pruning follows an iterative process – firstly, the importance of network elements is evaluated and the least important network elements are pruned away. Then, the model is fine-tuned on data to adapt to the pruned elements. This process repeats until the desired percentage of parameters are pruned away. A popular parameter importance criterion is by magnitude (magnitude-based pruning), in which parameters with smaller magnitudes are considered less substantial to the model's output and set to zero.

- In quantization, parameters are stored at a lower precision (usually, 8-bit integer form). This significantly decreases the storage and latency of a quantized model. However, performing post-processing quantization leads to accumulated inaccuracies that result in a large decrease in model performance. To address this, quantized models first undergo Quantization Aware Training, in which the model is in a simulated quantized environment and learns weights that are robust to quantization.

- In weight clustering, weights assigned to a cluster and set to the centroid value of that cluster, such that weights similar in value to one another (i.e., part of the same cluster) make slight adjustments to be the same. This redundancy in values allows for more efficient storage. The outcome of weight clustering is heavily dependent on the number of clusters chosen.

- In collaborative optimization, several model compression methods are chained together. By attaching model compression methods together, we can take advantage of each method's unique compression strengths. However, these methods must be attached in an order and implemented with special consideration to preserve the compression effect of the previous method.

Model compression methods can be implemented using the TensorFlow Model Optimization library. To implement model compression methods, use appropriate TensorFlow Model Optimization functions to wrap an existing Keras model in "prunable," "quantizable," or "clusterable" layers. After the model compression is performed, remove the compression wrappers from these layers. Often, you will need to apply a compression algorithm (e.g., GZIP) and convert the model into TFLite to see the rest of compression.

- Model compression (primarily pruning) forces models to sacrifice understanding of the long tail end of the data distribution, shrinking model generalization capability. It also increases compressed models' vulnerability to adversarial attacks and poses questions of fairness.

In the next chapter, we will discuss the automation of deep learning design with meta-optimization.

CHAPTER 5

Automating Model Design with Meta-optimization

Learning how to learn is life's most important skill.

—Tony Buzan, Writer and Educational Consultant

As the content we find we need to learn changes throughout life, we find new methods of learning that work for us. Through your education, you might have considered brute-force solving dozens of repetitive, minimally varying problems to be helpful when mastering the basics of algebra, active reading with a highlighter and notes on hand to help you succeed in more advanced English classes, and, later on with more advanced topics, understanding the conceptual framework and intuition more helpful than rotely solving a series of problems.

Ultimately, the task of learning is not just one of optimizing your mastery of the content within certain learning conditions but of optimizing those very learning conditions. To become efficient agents and designers of learning processes, we must recognize that learning is multi-tiered, controlled not only by our progress within the learning framework an agent may currently be operating in but also the learning framework itself.

This necessity applies to neural network design. Designing neural networks involves making many choices, and many of these can often feel arbitrary and therefore unoptimized or optimizable. While intuition is certainly a valuable guide in building model architectures, there are many aspects of neural network design that a human designer simply cannot effectively tune by hand, especially when multiple variables are involved.

© Andre Ye 2022
A. Ye, *Modern Deep Learning Design and Application Development*,
https://doi.org/10.1007/978-1-4842-7413-2_5

Meta-optimization, also referred to in this context as meta-learning or auto-ML, is the process of "learning how to learn" – a meta-model (or "*controller* model") finds the best parameters for the *controlled* model's validation performance. With its tools and knowledge about its underlying dynamics and best use cases, meta-optimization is a valuable box of methods to aid you in developing more structured, efficient models.

Introduction to Meta-optimization

A deep learning model is already a learner itself, optimizing its designated weights to maximize its performance on the training data it is given. *Meta-optimization* involves another model on a higher level optimizing the "fixed" parameters in the first model such that the first model, when trained under the conditions of those fixed parameters, will learn weights that maximize its performance on the test dataset (Figure 5-1). In machine learning – a domain in which meta-optimization is frequently applied – these fixed parameters could be factors like the gamma parameter in support vector classifiers, the value of k in the k-nearest neighbors algorithm, or the number of trees in a gradient boosting model. In the context of deep learning – the focus of this book and thus the application of meta-optimization in this chapter – these include factors like the architecture of the model and elements of its training procedure, like the choice of optimizer or the learning rate.

Note   Here, we use the terms "parameters" and "weights" selectively for the sake of clarity, even though the two are more or less synonymous. "Parameter" will refer to broader factors in the model's fundamental structure that remain unchanged during training and influence the outcome of the learned knowledge. "Weight" refers to changeable, trainable values that are used in representing a model's learned knowledge.

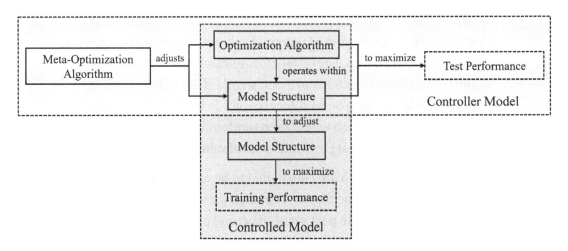

Figure 5-1. *Relationship between controller model and controlled model in meta-optimization*

So-called "naïve" meta-optimization algorithms use the following general structure:

1. Select structural parameters for a proposed controlled model.

2. Obtain the performance of a controlled model trained under those selected structural parameters.

3. Repeat.

There are two generally recognized naïve meta-optimization algorithms used as a baseline against more sophisticated meta-optimization methods:

- *Grid search*: In a grid search, every combination of a user-specified list of values for each parameter is tried and evaluated. Consider a hypothetical model with two structural parameters we would like to optimize, *A* and *B*. The user may specify the search space for *A* to be [1, 2, 3] and the search space for *B* to be [0.5, 1.2]. Here, "search space" indicates the values for each parameter that will be tested. A grid search would train six models for every combination of these parameters – $A = 1$ and $B = 0.5$, $A = 1$ and $B = 1.2$, $A = 2$ and $B = 0.5$, and so on.

- *Random search*: In a random search, the user provides information about a feasible distribution of potential values that each structural parameter could take on. For instance, the search space for *A* may be a normal distribution with mean 2 and standard deviation 1, and the search space for *B* might be a uniform choice from the list of values [0.5, 1.2]. A random search would then randomly sample parameter values and return the best performing set of values.

Grid search and random search are considered to be naïve search algorithms because they do not incorporate the results from their previously selected structural parameters into how they select the next set of structural parameters; they simply repeatedly "query" structural parameters blindly and return the best performing set. While grid and random search have their place in certain meta-optimization problems – grid search suffices for small meta-optimization problems, and random search proves to be a surprisingly strong strategy for models that are relatively cheaper to train – they cannot produce consistently strong results for more complex models, like neural networks. The problem is not that these naïve methods inherently cannot produce good sets of parameters, but that they take too long to do so.

A key component of the unique character of meta-optimization that distinguishes it from other fields of optimization problems is the impact of the evaluation step in amplifying any inefficiencies in the meta-optimization system. Generally, to quantify how good certain selected structural parameters are, a model is fully trained under those structural parameters and its performance on the test set is used as the evaluation. (See the "Neural Architecture Search" section for proxy evaluation to learn about faster alternatives). In the context of neural networks, this evaluation step can take hours. Thus, an effective meta-optimization system should attempt to require as few models to be built and trained as possible before arriving at a good solution. (Compare this to standard neural network optimization, in which the model queries the loss function and updates its weights accordingly anywhere from hundreds of thousands to millions of times in the span of hours.)

To prevent inefficiency in the selection of new structural parameters to evaluate, successful meta-optimization methods used for models like neural networks include another step – incorporating knowledge from previous "experiments" into determining the next best set of parameters to select:

1. Select structural parameters for a proposed controlled model.

2. Obtain the performance of a controlled model trained under those selected structural parameters.

3. Incorporate knowledge about the relationship between selected structural parameters and the performance of a model trained under such parameters into the next selection.

4. Repeat.

Even with these adaptations, meta-optimization methods are taxing on computational and time resources. A primary factor in the success of a meta-optimization campaign is how you define the *search space* – the feasible distribution of values the meta-optimization algorithm can draw from. Choosing the search space is another trade-off. Clearly, if you specify too large a search space, the meta-optimization algorithm will need to select and evaluate more structural parameters to arrive at a good solution. Each additional parameter enlarges the existing search space by a significant factor, so leaving too many parameters to be optimized by the meta-optimization algorithm will likely perform worse than user-specified values or a random search, which doesn't need to deal with the complexities of navigating an incredibly sparse space.

Herein lies an important principle in meta-optimization design: *be conservative in determining parameters to be optimized by meta-optimization as possible.* If you know that batch normalization will benefit a network's performance, for instance, it's probably not worth it to use meta-optimization to determine if batch normalization should be included in the network architecture or not. Moreover, if you decide a certain parameter should be optimized via meta-optimization, attempt to decrease its "size." For instance, this could be the number or range of possible values a parameter can take on, the range of possible values.

On the other hand, if you define too small a search space, you should ask yourself another question – *is meta-optimization worth performing in the first place*? A meta-optimization algorithm is likely to find the optimal set of parameters for a search space defined as {A: normal distribution with mean 1 and standard deviation 0.001 and B: uniform distribution from 3.2 to 3.3} very efficiently, for instance, but it's not useful. The user could have likely set $A=1$ and $B=3.25$ with no visible impact on the resulting model's performance.

> **Note**   What is a "small" or "large" range is dependent on the nature of the
> parameter and the variation required to make visible change in the performance of
> the model. Parameters sampled from a normal distribution with mean 0.005 and
> standard deviation 0.001 may yield a very similar model if that parameter is the C
> parameter in a support vector machine. However, if the parameter is the learning
> rate of a deep learning model, it is likely that such a distribution would yield visible
> differences in model test performance.

Thus, the crucial balance in meta-optimization is that of engineering a search space conservative enough not to be redundant, but free and "open" enough to yield significant results.

This chapter will discuss two forms of meta-optimization as applicable to deep learning: general hyperparameter optimization and Neural Architecture Search (NAS), along with the Hyperopt, Hyperas, and Auto-Keras libraries.

General Hyperparameter Optimization

General hyperparameter optimization is a broad field within meta-optimization concerning general methods to optimize the parameters of a wide variety of models. These methods are not explicitly built for neural network designs, so additional work will be required for effective results.

In this section, we'll discuss Bayesian optimization – the leading general hyperparameter optimization method for machine and deep learning, as well as the usage of the popular meta-optimization library Hyperopt and its accompanying Keras wrapper, Hyperas, to optimize neural network design.

Bayesian Optimization Intuition and Theory

Here's a function: $f(x)$. You only have access to its output given a certain input, and you know that it is expensive to calculate. Your task is to find the set of inputs that minimizes the output of the function as much as possible.

This sort of setup is known as a black-box optimization problem, because the algorithm or entity attempting to find a solution to the problem has access to very little information about the function (Figure 5-2). You have access only to the output of any input passed into the function, but not the derivative, barring the usage of gradient-based methods that have proved successful in the domain of neural networks. Moreover, because $f(x)$ is expensive to evaluate (i.e., takes a significant amount of time to get the output of an input), we cannot employ a host of non-gradient optimization methods from the simple grid search to the more sophisticated simulated annealing. These methods require a large quantity of queries to the black-box function to discover reasonably well-performing results.

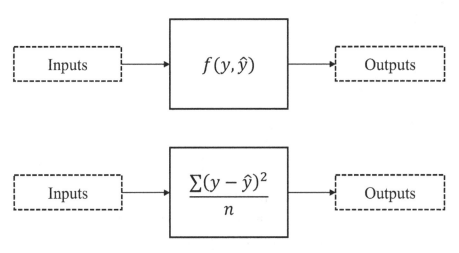

Figure 5-2. *Objective functions to minimize. Top – black-box function; bottom – explicitly defined loss function (in this case MSE)*

Bayesian optimization is often used in these sorts of black-box optimization problems because it succeeds in obtaining reliably good results with a relatively small number of required queries to the objective function $f(x)$. Hyperopt, in addition to many other libraries, uses optimization algorithms built upon the fundamental model of Bayesian optimization. The spirit of Bayesian modeling is to begin with a set of *prior* beliefs and continually update that set of beliefs with new information to form posterior beliefs. It is this spirit of continuous update – searching for new information in places where it is needed – that makes Bayesian optimization a robust and versatile tool in black-box optimization problems.

Consider this hypothetical objective function, $c(x)$ (Figure 5-3). In the context of meta-learning/meta-optimization, $c(x)$ represents the loss or cost of a certain model and x represents the set of parameters used in the model that are being optimized.

Figure 5-3. *Hypothetical cost function – the loss incurred by a model with some parameter x. For the sake of visualization, in this case we are optimizing only one parameter*

Because this is a black-box function and the Bayesian optimization algorithm doesn't "know" its full shape, it develops its own representations of what it "thinks" the objective function looks like via a *surrogate function* (Figure 5-4). The surrogate function approximates the objective function and represents the current set of beliefs on how the objective function behaves.

Note that the representation of the surrogate model here is deterministic, but in practice it is a probabilistic model that returns $p(y|x)$ or the probability that the objective function's output is y given an input x. Probabilistic surrogate models are easier and more natural to update in a Bayesian function.

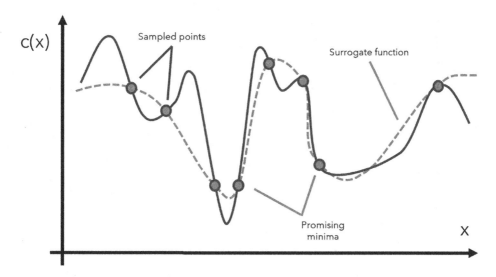

Figure 5-4. *Example surrogate function and how the surrogate function informs the sampling of new points in the objective function*

Based on the surrogate function, the algorithm can identify which points look "promising" in addition to which areas need more exploration and samples from these promising regions accordingly (Figure 5-5). Note that there is an exploration-exploitation trade-off dynamic at play here: if the algorithm sampled only from regions immediately suggested to be minima (purely exploiting), it would completely overlook other promising minima that were not captured in the first round of sampling. Likewise, if the algorithm was purely explorative, it would behave little differently from a random search by not considering previous findings. The *acquisition function* is responsible for determining how surrogate functions are updated.

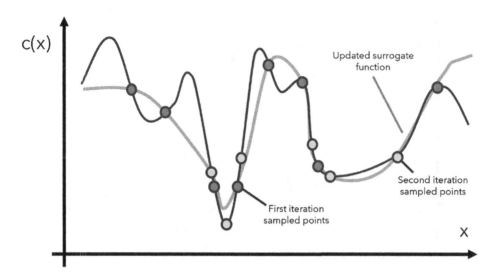

Figure 5-5. *Updating the surrogate function with the new, second iteration of sampled points*

After just a few number of iterations, there is a very high probability that the Bayesian optimization algorithm has obtained a relatively accurate representation of the minima within the black-box function.

Because a random or grid search does not take any of the previous results into consideration when determining the next sampled set of parameters, these "naïve" algorithms save time in calculating the next set of parameters to sample. However, the additional computation Bayesian optimization algorithms use to determine the next point to sample is used to construct a surrogate function more intelligently with fewer queries. Net-wise, the reduction in necessary queries to the objective function outweighs the increase in time to determine the next sampled point, making the Bayesian optimization method more efficient.

This process of optimization is known more abstractly as *Sequential Model-Based Optimization (SMBO)*. It operates as a central concept or template against which various model optimization strategies can be formulated and compared and contains one key feature: a surrogate function for the objective function that is updated with new information and used to determine new points to sample. Across various SMBO methods, the primary differentiators are the design of the acquisition function and the method of constructing the surrogate model. Hyperopt uses the Tree-structured Parzen Estimator (TPE) surrogate model and acquisition strategy.

The expected improvement measurement quantifies the expected improvement with respect to the set of parameters to be optimized, x. For instance, if the surrogate model $p(y|x)$ evaluates to zero for all values of y less than some threshold value $y*$ – that is, there is zero probability that the set of input parameters x could yield an output of the objective function less than $y*$ – there is likely little improvement.

The Tree-structured Parzen Estimator (Figure 5-6) is built to work toward a set of parameters x that maximizes expected improvement. Like all surrogate functions used in Bayesian optimization, it returns $p(y|x)$ – the probability that the objective function's output is y given an input x. Instead of directly obtaining this probability, it uses Bayes' rule:

$$p(x) = \frac{p(y) \cdot p(y)}{p(x)}$$

The $p(x|y)$ term represents the probability that the input to the objective function was x given an output y. To calculate this, two distribution functions are used: $l(x)$ if the output y is less than some threshold $y*$ and $g(x)$ if the output y is less than some threshold $y*$. To sample values of x that yield objective function outputs less than the threshold, the strategy is to draw from $l(x)$ rather than $g(x)$. (The other terms, $p(y)$ and $p(x)$, can be easily calculated as they do not involve conditionals.) Sampled values with the highest expected improvement are evaluated through the objective function. The resulting value is used to update the probability distributions $l(x)$ and $g(x)$ for better prediction.

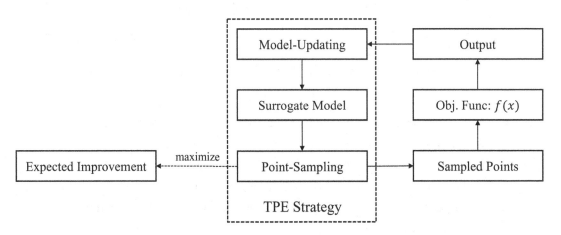

Figure 5-6. *Visualizing where the Tree-structured Parzen Estimator strategy falls in relation to the Sequential Model-Based Optimization procedure*

It's a little bit math heavy, but ultimately the Tree-structured Parzen Estimator strategy attempts to find the best objective function inputs to sample by continually updating its two internal probability distributions to maximize the quality of prediction.

Note   You may be wondering – what is tree-structured about the Tree-structured Parzen Estimator strategy? In the original TPE paper, the authors suggest that the "tree" component of the algorithm's name is derived from the tree-like nature of the hyperparameter space: the value chosen for one hyperparameter determines the set of possible values for other parameters. For instance, if we are optimizing the architecture of a neural network, we first determine the number of layers before determining the number of nodes in the third layer.

Hyperopt Syntax, Concepts, and Usage

Hyperopt is a popular framework for Bayesian optimization. Its flexible syntax allows for you to test hyperparameter tuning on any framework and for any purpose. Using Hyperopt requires three key elements (Figure 5-7):

- *Objective function*: This is a function that takes in a dictionary of hyperparameters (the inputs to the objective function) and outputs the "goodness" of those hyperparameters (the output of the objective function). In this context of meta-learning, the objective function takes in the hyperparameters, uses the hyperparameters to build a model, trains the model, and returns the performance of the model on validation/test data. "Better" is synonymous with "less" in Hyperopt, so make sure that metrics like accuracy are negated.

- *Search space*: This is the space of parameters with which Hyperopt will search. It is implemented as a dictionary of parameters in which the key is the name of the parameter (for your own reference later) and the corresponding value is a Hyperopt search space objective defining the range and type of distribution to sample values for that parameter from.

- *Search*: Once you have defined the objective function and the search space, you can initiate the actual search function, which will return the best set of parameters from the search.

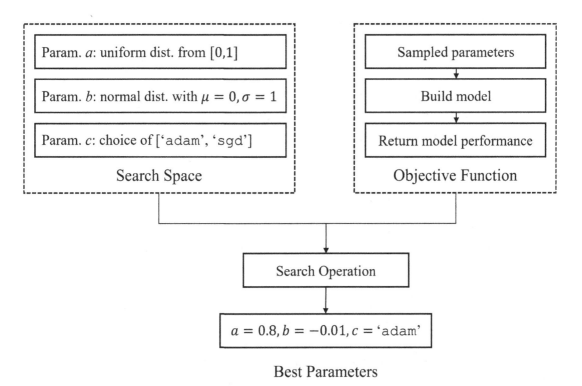

Figure 5-7. *Relationship between the search space, objective function, and search operation in the Hyperopt framework*

Install Hyperopt with `pip install hyperopt` and import with `import hyperopt`.

Hyperopt Syntax Overview: Finding the Minimum of a Simple Objective Function

To illustrate the basic syntax and concepts in Hyperopt, we will use Hyperopt to solve a very simple problem: find the minimum of $f(x) = (x-1)^2$. Let's first define the search space with `hyperopt.hp` (Listing 5-1).

Listing 5-1. Importing the Hyperopt library and defining the search space for a single parameter using hyperopt.hp

```
from hyperopt import hp
space = {'x':hp.uniform('x',-1000,1000)}
```

In this case, we are telling Hyperopt that the search space consists of one parameter with label "x", which can reasonably be found from a uniform distribution from –1000 to 1000. However, from domain knowledge we know that it's more likely that the optimal value of x that minimizes the objective function is more likely to be near zero than equally likely to be any value from –1000 to 1000. Ideally, we would like the optimizer to sample values of x near zero more often than values near 1000 or –1000. We can formulate this domain knowledge regarding the probability that the optimal value is near a certain value by using other spaces (see Figure 5-8):

- hp.normal(label, mu, sigma): The distribution of the search space for this parameter is a normal distribution with mean mu and standard deviation sigma.

- hp.lognormal(label, mu, sigma): The distribution of the search space for this parameter is a log-normal distribution with mean mu and standard deviation sigma. It acts as a modification of hp.normal that returns only positive values. This is useful for parameters like the learning rate of a neural network that are continuous and contain a concentration of likelihood at some point, but require a positive value.

- hp.qnormal(label, mu, sigma, q) and hp.qlognormal(label, mu, sigma, q): These act as distributions for quasi-continuous parameters, like the number of layers or number of nodes within a layer in a neural network – while these are not continuous (a network with 3.5 layers is invalid), they contain transitive relative relationships (a network with 4 layers is longer than a network with 3 layers). Correspondingly, we may want to formulate certain relationships, like wanting the number of layers to be shorter than longer. hp.qnormal and hp.qlognormal "quantize" the outputs of hp.normal and hp.lognormal by performing the following operation: $q \cdot \mathrm{round}\left(\dfrac{o}{q}\right)$, where o is the output of the "unquantized" operation and q is the quantization factor. hp.qnormal('x', 5, 3, 1), for instance, defines a search space of "normally distributed" integers ($q = 1$) with mean 5 and standard deviation 3.

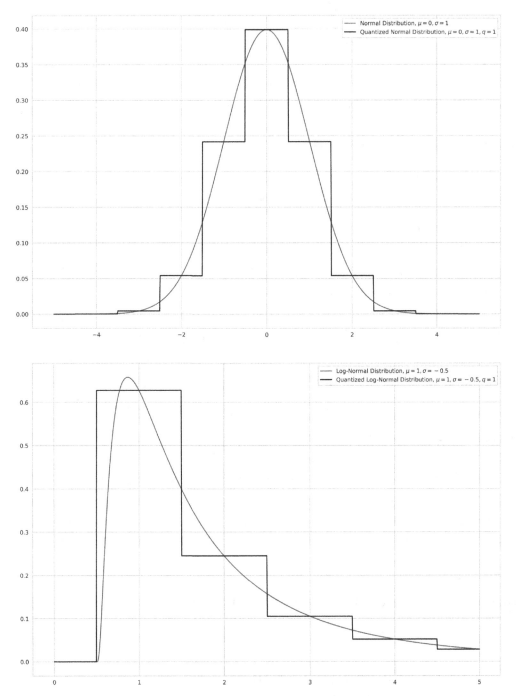

Figure 5-8. *Visualizations of the normal and quantized normal distributions (top) and the log-normal and quantized log-normal distributions (bottom). The quantized distribution visualization is not completely faithful – values that fall into a sampled "segment" are assigned the same value*

If the search space is not continuous or quasi-continuous, but instead a series of discrete, non-comparable choices, use hp.choice(), which takes in a list of possible choices.

In this case, we can use a standard normal distribution to more accurately describe where we think the optimal location of the parameter is (Listing 5-2).

Listing 5-2. Defining a simple Hyperopt space

```
from hyperopt import hp
space = {'x':hp.normal('x', mu=0, sigma=10)}
```

We can now define the objective function, which simply involves subtracting 1 from x and squaring: obj_func = lambda params: (params['x']-1)**2.

Only returning the associated output of the objective function is fine if the objective function always returns a valid output or you have configured the search space such that it is impossible for any invalid input to be passed into the objective function. If not, however, Hyperopt provides one additional feature that may be helpful if certain combinations of parameters may be invalid: a status. For instance, if we are trying to find the minimum of $f(x) = \left|\dfrac{1}{x}\right| + x^2$, the input $x = 0$ would be invalid. There's no easy way to restrict the search space to exclude $x = 0$, though. If $x \neq 0$, the output of the objective function is {'loss':value, 'status':'ok'}. If the input parameter is equal to 0, though, the objective function returns {'status':'fail'}.

In the context of modeling, certain parameter constructions may not be valid. Given that restricting the search space is impossible or too difficult, you can construct your objective function with a try/except catching mechanism such that any error Keras throws when building the graph is communicated to Hyperopt in the form of a failed status (Listing 5-3).

Listing 5-3. An objective function with ok/fail status

```
def obj_func(params):
    try:
        model = build_model().train(data)
        loss = evaluate(model)
        return {'loss':loss, 'status':'ok'}
    except:
        return {'status':'fail'}
```

Now that the search space and the objective function have been defined, we can initiate a search to find the parameter values specified in the search space that minimize the objective function (Listing 5-4).

Listing 5-4. Hyperopt minimization procedure

```
from hyperopt import fmin, tpe
best = fmin(obj_func, space, algo=tpe.suggest, max_evals=500)
```

Here, `algo=tpe.suggest` uses the Tree-structured Parzen Estimator optimization algorithm and `max_evals=500` lets Hyperopt know that the code will tolerate a maximum 500 evaluations of the objective function. In the context of modeling, `max_evals` indicates the maximum number of models that will be built and trained, because each evaluation of the objective function requires building a new model architecture, training it, evaluating it, and returning its performance.

After the search completes, `best` is a dictionary of the best parameters found. `best['x']` should contain a value very close to 1 (the true minimum).

Using Hyperopt to Optimize Training Procedure

Parameters involved in a model's training procedure include the learning rate, the choice of optimizer, callbacks, and other parameters involved in how the model is trained rather than the architecture of the model. Let's use Hyperopt to determine the optimal optimizer and learning rate for training. We'll need to define specific search space types for these two parameters:

- `hp.choice('optimizer', ['adam', 'rmsprop', 'sgd'])` for the optimizer: This will find the optimal optimizer to train the network on.

- `hp.lognormal('lr', mu=0.005, sigma=0.001)` for the optimizer learning rate: The log-normal distribution is used here because the learning rate must be positive.

We can define these two spaces in a dictionary (Listing 5-5). Note that we import optimizer objects without actually instantiating them (i.e., `keras.optimizers.Adam` rather than `keras.optimizers.Adam()`). This is because we need to pass the learning rate as a parameter within the instantiation of the parameter object, which we'll do when we're building the model in the objective function.

Listing 5-5. Defining a search space for model optimizer and learning rate

```
from keras.optimizers import Adam, RMSprop, SGD
optimizers = [Adam, RMSprop, SGD]
space = {'optimizer':hp.choice('optimizer',optimizers),
         'lr':hp.lognormal('lr', mu=0.005, sigma=0.001)}
```

The objective function will take in a dictionary of parameters sampled from the search space and use them to train a model architecture (Listing 5-6). In this particular case, we will measure model performance by its accuracy on the test dataset. We perform the following operations in the objective function:

1. *Build the model*: We will use a simple sequential model with a convolutional and fully connected component. This can be built without accessing the `params` dictionary because we're not tuning any hyperparameters that influence how the architecture is built.

2. *Compile*: This is the relevant component of the construction and training of the model because parameters we are tuning (optimizer and learning rate) are explicitly used in this step. We will instantiate the sampled optimizer with the sampled learning rate and then pass the optimizer object with that learning rate into `model.compile()`. We will also pass `metrics=['accuracy']` into compiling such that we can access the accuracy of the model on the test data in evaluation as the output of the objective function.

3. *Fit model*: We fit the model as usual for some number of epochs.

4. *Evaluate accuracy*: We can call `model.evaluate()` to return a list of loss and metrics calculated on the test data. The first element is the loss, and the second is the accuracy; we index the output of evaluation accordingly to assess the accuracy.

5. *Return negated accuracy on validation set*: The accuracy is negated such that smaller is "better."

Listing 5-6. Defining the objective function of a training procedure-optimizing operation

```python
from keras.models import Sequential
import keras.layers as L

def objective(params):
    # build model
    model = Sequential()
    model.add(L.Input((32,32,3)))
    for i in range(4):
        model.add(L.Conv2D(32, (3,3), activation='relu'))
    model.add(L.Flatten())
    model.add(L.Dense(64, activation='relu'))
    model.add(L.Dense(1, activation='sigmoid'))

    # compile
    optimizer = params['optimizer'](lr=params['lr'])
    model.compile(loss='binary_crossentropy',
                  optimizer=optimizer,
                  metrics=['accuracy'])

    # fit
    model.fit(x_train, y_train, epochs=20, verbose=0)

    # evaluate accuracy (second elem. w/ .evaluate())
    acc = model.evaluate(x_test, y_test, verbose=0)[1]

    # return negative of acc such that smaller = better
    return -acc
```

Note that we specify verbose=0 with both model.fit() and model.evaluate(), which prevents Keras from printing progress bars and metrics during training. While these progress bars are helpful when training a Keras model in isolation, in the context of hyperparameter optimization with Hyperopt, they interfere with Hyperopt's own progress bar printing.

We can use the objective function and the search space, along with the Tree-structured Parzen Estimator and the maximum number of evaluations, into the `fmin` function: best = fmin(objective, space, algo=tpe.suggest, max_evals=30).

After the search has completed, `best` should contain a dictionary of best performing values for each parameter specified in the search space. In order to use these best parameters in your model, you can rebuild the model as it is built in the objective function and replace the `params` dictionary with the `best` dictionary.

For even better performance, two adaptations should be made:

- *Early stopping callback*: This stops training after performance plateaus to save as many resources (computational and time-wise) as possible, as meta-optimization is an inherently expensive operation. This can usually be coupled with a high number of epochs, such that each model design is trained toward fruition – its potential is extracted as much as reasonably possible, and training stops after there seems to be no more potential to extract.

- *Model checkpoint callback with weight reloading before evaluation*: Rather than evaluating the state of the neural network after it has completed training, it is optimal to evaluate the best "version" of that neural network. This can be done via the model checkpoint callback, which saves the weights of the best performing model. Before evaluating the performance of the model, reload these weights.

These two measures will further increase the efficiency of the search.

Using Hyperopt to Optimize Model Architecture

Although Hyperopt is often used to tune parameters in the model's training procedure, you can also use it to make fine-tuned optimizations to the model architecture. It's important, though, to consider whether you should use a general meta-optimization method like Hyperopt or a more specialized architecture optimization method like Neural Architecture Search. If you want to optimize large changes in the model architecture, it's best to use Neural Architecture Search via Auto-Keras (this is covered later in the chapter). On the other hand, if you want to optimize small changes, Auto-Keras may not offer you the level of precision you desire, and thus Hyperopt may be the better solution. Note that if the change in architecture you intend to optimize is very small (like finding the optimal number of neurons in a layer), it may not be fruitful to even optimize it at all, provided that you have set a reasonable default parameter.

Good architecture components to optimize with a general optimization framework like Hyperopt that are neither too large to optimize with a more specialized Neural Architecture Search method nor too small to be insignificant with respect to the model performance include

- *Number of layers in a certain block/component* (provided the range is long enough): The number of layers is quite a significant factor in the model architecture, especially if it is compounded via a block/cell-based design.

- *Presence of batch normalization*: Batch normalization is an important layer that aids in smoothing the loss space. However, it succeeds only if used in certain locations and with a certain frequency.

- *Presence and rate of dropout layer*: Like batch normalization, dropout can be an incredibly powerful regularization method. Successful usage of dropout requires placement in certain locations, with a certain frequency, and a well-tuned dropout rate.

For this example, we'll tune three general factors of the model architecture: the number of layers in the convolutional component, the number of layers in the dense component, and the dropout rate of a dropout layer inserted in between every layer. (You could also tune the dropout rate of all dropout layers, which offers less customizability but may be more successful in some circumstances.)

Because we are keeping track of the dropout rate of several dropout layers, we cannot merely define it as a single parameter in the search space. Rather, we will need to automate the storage and organization of several parameters in the search space corresponding to each dropout layer.

Let's begin by defining some key variables that we'll need to use repeatedly later (Listing 5-7). We're willing to accept any number of convolutional layers from a minimum of 3 layers to a maximum of 8 layers and any number of dense layers from 2 to 5. (You can adjust this, of course, to your particular problem.)

Listing 5-7. Defining key parameters

```
min_num_convs = 3
max_num_convs = 8
min_num_dense = 2
max_num_dense = 5
```

Using this information, we will generate two lists, `conv_drs` and `dense_drs`, which contain a Hyperopt search space object for the dropout rate of each layer in the convolutional and dense components, respectively (Listing 5-8). This allows us to store multiple related but different search parameters efficiently; these can be accessed easily via indexing when constructing the model. We use string formatting to provide a unique string name to the Hyperopt search space. Note that while the name you provide to each search space parameter is arbitrary (the user accesses each parameter through other means), Hyperopt requires (a) string names and (b) unique names (i.e., no two parameters can have the same name).

Listing 5-8. Creating an organized list of dropout rates

```
conv_drs, dense_drs = [], []
for layer in range(max_num_convs):
    conv_drs.append(hp.normal(f'c{layer}', 0.15, 0.1))
for layer in range(max_num_dense):
    dense_drs.append(hp.normal(f'd{layer}', 0.2, 0.1))
```

Note that we are constructing as many search space variables as the maximum number of layers for the convolutional and fully connected components, which means that there will be redundancy if the number of layers sampled is less than the maximum number (i.e., some dropout rates will not be used). This is fine; Hyperopt will handle it and adapt to these relationships. Additionally, we are defining a normal search space for the dropout rate that could theoretically sample values less than 0 or larger than 1 (i.e., invalid dropout rates). We will adjust these within the objective function to demonstrate custom manipulation of the search space when Hyperopt does not provide a function that fits your particular needs (in this case, a normal-shaped distribution that is bounded on both ends).

We can use these parameters to create the search space (Listing 5-9). When defining the number of layers in the convolutional and dense components, we use a quantized uniform distribution with q=1 to sample all integers from min_num_convs/dense to max_num_convs/dense (inclusive). Additionally, note that we passed in lists for the 'conv_dr' and 'dense_dr' parameters. Hyperopt will interpret this (or any other data type that contains several Hyperopt search space objects) as a sub-class of parameters that will be tuned like any other parameter.

Listing 5-9. Defining the search space for optimizing neural network architecture

```
space = {'#convs':hp.quniform('#convs',
                              min_num_convs,
                              max_num_convs,
                              q=1),
        '#dense':hp.quniform('#dense',
                              min_num_dense,
                              max_num_dense,
                              q=1),
        'conv_dr':conv_drs,
        'dense_dr':dense_drs}
```

Building the objective function in this context is an elaborate process, so we'll build it in multiple pieces.

Recall that the Hyperopt search space for the dropout rate was defined to be normally distributed, meaning that it is possible to sample invalid dropout rates (less than 0 or larger than 1). We can address sampled parameters that are invalid at the beginning of the objective function (Listing 5-10).

If the dropout rate is larger than 0.9, we set it to 0.9 (Keras does not accept a dropout rate equal to 1, and any dropout rate larger than 90% is unlikely to succeed anyways). On the other hand, if the dropout rate is less than 0, we set it to 0. Given the mean and standard deviation parameters defined in the search space, it is unlikely for either of these to be sampled, but it's important to define these catch mechanisms to prevent errors that disrupt the optimization process. Note that another alternative would be to return {'status':'fail'} to indicate an invalid parameter(s). The Bayesian optimization algorithm will adapt to any of these measures.

Listing 5-10. Beginning to define the objective function – correcting for dropout rates sampled in an invalid domain

```
def objective(params):

    # convert set of params to list for mutability
    conv_drs = list(params['conv_dr'])
    dense_drs = list(params['dense_dr'])

    # make sure dropout rate is 0 <= r < 1
    for ind in range(len(conv_drs)):
        if conv_drs[ind] > 0.9:
            conv_drs[ind] = 0.9
        if conv_drs[ind] < 0:
            conv_drs[ind] = 0
    for ind in range(len(dense_drs)):
        if dense_drs[ind] > 0.9:
            dense_drs[ind] = 0.9
        if dense_drs[ind] < 0:
            dense_drs[ind] = 0
    ...
```

We can then build the model "template" (the Sequential base model) and attach an input to it (Listing 5-11).

Listing 5-11. Defining the model template and input in the objective function

```
...
# build model template + input
model = Sequential()
model.add(L.Input((32,32,3)))
...
```

When building the convolutional component, we add however many convolutional layers specified in the input parameters via a for loop (Listing 5-12). Note that we wrap the sampled number of convolutional layers params['#convs'] in an int() function; the output of the quantized value will not technically be an integer (e.g., 3.0, 4.0), whereas Python requires an integer object input to the range() function. After each convolutional layer, we add a dropout layer with the dropout rate accessed by indexing the previously defined conv_drs list of dropout rates. By organizing dropout rates into the easily accessible and storable list format, we are able to integrate several parameters into the optimization procedure.

Listing 5-12. Building the convolutional component in the objective function

```
...
# build convolutional component
for ind in range(int(params['#convs'])):

    # add convolutional layer
    model.add(L.Conv2D(32, (3,3), activation='relu'))

    # add corresponding dropout rate
    model.add(L.Dropout(conv_drs[ind]))

# add flattening for dense component
model.add(L.Flatten())
...
```

Constructing the dense component follows the same logic (Listing 5-13).

Listing 5-13. Building the dense component in the objective function

```
...
# build dense component
for ind in range(int(params['#dense'])):

    # add dense layer
    model.add(L.Dense(32, activation='relu'))

    # add corresponding dropout rate
    model.add(L.Dropout(dense_drs[ind]))
...
```

Afterward, append the model output and perform the previously discussed remaining steps of compiling, fitting, evaluating, and returning the output of the objective function.

As you can see, Hyperopt allows for a tremendous amount of control over specific elements of the model, even if it involves a little bit more work – your imagination (and your capability for organization) is the limit!

Hyperas Syntax, Concepts, and Usage

Hyperopt offers a tremendous amount of customizability and adaptability toward your particular optimization needs, but it can be a lot of work, especially for relatively simpler tasks. Hyperas is a wrapper that operates on Hyperopt but is specialized in syntax for meta-optimizing Keras models. The primary advantage of Hyperas is that you can define parameters to be optimized with much fewer code than Hyperopt syntax requires.

Note   While Hyperas is a useful resource, it's important to know how to use Hyperopt because (a) often, problems that require meta-optimization in the first place are complex enough to warrant using Hyperopt and (b) Hyperas is a less developed and stable package (it is currently archived by the owner). Additionally, be warned that there are complications with using Hyperas with Jupyter Notebooks in an environment like Kaggle or Colab – if you are working with these circumstances, it may be easier to use Hyperopt.

Install Hyperas with `pip install hyperas` and import with `import hyperas`.

Using Hyperas to Optimize Training Procedure

Let's use Hyperas' syntax to perform the same optimization of the model training procedure by finding the best combination of optimizer and learning rate. Hyperas has three primary components (Figure 5-9):

- *Data feeder function*: A function must be defined to load data, perform any preprocessing, and return four sets of data: x train, y train, x test, and y test (in that order). By defining the data feeding process as a function, Hyperas can make shortcuts to prevent redundancy in data loading in training the model.

- *Objective function*: This function takes in the four sets of data from the data feeder function. It builds the model with unique markers for parameters to be optimized, fits on the training data, and returns whatever loss is used for evaluation. The objective function should return a dictionary with three key-value pairs: the loss, the status, and the model the loss was evaluated on.

- *Minimization*: This function takes in the objective function (model creating function) and the data feeder function, alongside the Tree-structured Parzen Estimators algorithm from Hyperopt and a max_evals parameter. Because Hyperas is implemented with lots of recording, the minimization procedure requires a hyperopt. Trials() object, which serves as an additional documentation/ recording object. (You can also pass this into fmin() for Hyperopt, although it's not required.)

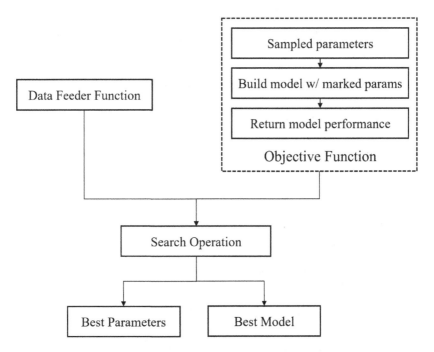

Figure 5-9. *Key components of the Hyperas framework – data feeder, objective function, and search operation*

Assuming the data feeder function has already been created, we will create the objective function (Listing 5-14). It is almost identical to the objective function in Hyperopt, with two key differences: the objective function takes in the four sets of data rather than the parameters dictionary, and optimizable parameters are defined completely within the objective function and with as little user attachment and handling as necessary.

To specify an optimizable parameter, put double braces around a `hyperas.distributions` search space distribution object (e.g., `{{hyperas.distributions.choice(['a','b','c'])}}`). Hyperas contains all the distributions implemented in Hyperopt. Note that no label is required, which allows for easy definition of numerous optimizable parameters. The double-brace syntax can only be used in the context of the objective function, which Hyperas uses Jinja style for template replacement and temporary files. Note that because Hyperas is creating these models in separate "environments," you may need to re-import certain models or layers again within the objective function. In this case, we import the Sequential model and the Keras layers.

Listing 5-14. Objective function for optimizing training procedure in Hyperas

```python
from hyperas.distributions import choice, lognormal
from keras.optimizers import Adam, RMSprop, SGD

def obj_func(x_train, y_train, x_test, y_test):

    # import keras layers and sequential model
    from keras.models import Sequential
    import keras.layers as L

    # define model
    model = Sequential()
    model.add(L.Input((32,32,3)))
    for i in range(4):
        model.add(L.Conv2D(32, (3,3), activation='relu'))
    model.add(L.Flatten())
    model.add(L.Dense(64, activation='relu'))
    model.add(L.Dense(1, activation='sigmoid'))
```

```
# sample lr and optimizer (not instantiated yet)
lr = {{lognormal(mu=0.005, sigma=0.001)}}
optimizer_obj = {{choice([Adam, RMSprop, SGD])}}

# instantiate sampled optimizer with sampled lr
optimizer = optimizer_obj(lr=lr)

# compile with sampled parameters
model.compile(loss='binary_crossentropy',
              optimizer=optimizer,
              metrics=['accuracy'])

# fit and evaluate
model.fit(x_train, y_train, epochs=1, verbose=0)
acc = model.evaluate(x_test, y_test, verbose=0)[1]

# return loss, OK status, and trained candidate model
return {'loss':-acc, 'status':'ok', 'model':model}
```

To perform optimization, use the hyperas.optim.minimize function (Listing 5-15), which helpfully returns both the best parameters and the best model from the optimization procedure. (Recall that Hyperopt returns only the best set of parameters, which you need to store into a rebuilt model.) optim.minimize() takes in the user-specified objective function and the data feeder function, as well as the tpe.suggest and Trials() entities from Hyperopt. If you are working in Jupyter Notebooks, optim.minimize() also requires the name of your notebook.

Listing 5-15. Optimizing in Hyperas

```
from hyperas import optim
from hyperopt import tpe, Trials
best_pms, best_model = optim.minimize(model=obj_func,
                                      data=data,
                                      algo=tpe.suggest,
                                      max_evals=5,
                                      trials=Trials(),
                                      notebook_name='name')
```

After training, you can save the model and parameters for later usage.

Using Hyperas to Optimize Model Architecture

The true convenience of Hyperas is exposed when applied to a task like earlier of optimizing a host of parameters all at once. Rather than needing to create elaborate lists and organization structures for the search space, we can define parameters to optimize within the function itself (Listing 5-16). To prevent sampled parameters for the dropout rate from exceeding 0.9 or falling below 0, we can implement a custom rounding function, r, which takes in a parameter x_ (an underscore was added to distinguish it from x, which Hyperas uses internally and can cause problems) and either adjusts it if it is invalid or lets it pass. We wrap r around all sampled rates.

Listing 5-16. Objective function for optimizing architecture in Hyperas

```
def obj_func(x_train, y_train, x_test, y_test):

    # create rounding function
    import keras.layers as L
    r = lambda x_: 0 if x_<0 else (0.9 if x_>0.9 else x_)

    # import keras layers and sequential model
    from keras.models import Sequential
    import keras.layers as L

    # create model template and input
    model = keras.models.Sequential()
    model.add(L.Input((32,32,3)))

    # build convolutional component
    for ind in range(int({{quniform(3,8,1)}})):
        model.add(L.Conv2D(32, (3,3), activation='relu'))
        model.add(L.Dropout(r({{normal(0.2,0.1)}})))

    # add flattening layer for FC component
    model.add(L.Flatten())

    # build FC component
    for ind in range(int({{quniform(2,5,1)}})):
        model.add(L.Dense(32, activation='relu'))
        model.add(L.Dropout(r({{normal(0.2,0.1)}})))
```

```
# add output layer
model.add(L.Dense(1, activation='sigmoid'))

# compile, fit, evaluate, and return
model.compile(loss='binary_crossentropy',
              optimizer='adam',
              metrics=['accuracy'])
model.fit(x_train, y_train, epochs=1, verbose=0)
acc = model.evaluate(x_test, y_test, verbose=0)[1]
return {'loss':-acc, 'status':'ok', 'model':model}
```

This objective function can then be used in `hyperas.optim.minimize()` as usual.

Making adaptations is simple too – if you want to have one dropout rate per component, for instance, just sample the dropout rate outside of the loop such that only one parameter is created for the entire component (Listing 5-17).

Listing 5-17. The same dropout rate is used in every added layer by defining only one dropout rate that is repeatedly used in each dropout layer

```
conv_comp_rate = r({{normal(0.2,0.1)}})
for ind in range(int({{quniform(3,8,1)}})):
    model.add(L.Conv2D(32, (3,3), activation='relu'))
    model.add(L.Dropout(conv_comp_rate))
```

Neural Architecture Search

In our previous discussion of meta-optimization methods, we have used generalized frameworks for the optimization of parameters in all sorts of contexts that are also applicable to neural network architecture and training procedure optimization. While Bayesian optimization suffices for some relatively detailed or non-architectural parameter optimizations, in other cases we desire a meta-optimization method that is designed particularly for the task of optimizing the architecture of a neural network.

Neural Architecture Search (NAS) is the process of automating the engineering of neural network architectures. Because NAS methods are designed specifically for the task of searching architectures, they are generally more efficient in finding high-performing architectures than more generalized optimization methods like Bayesian optimization. Moreover, because Neural Architecture Search often requires searching

for practical, efficient architectures, many view NAS as a form of model compression –
building an architecture that can effectively represent the same knowledge as a larger
architecture more effectively.

NAS Intuition and Theory

Many of the well-known deep learning architectures – ResNet and Inception, for
instance – are built with incredibly complex structures that require a team of deep
learning engineers to conceive and experiment with. The process of building such
structures, too, has never quite been a precise science, but instead a continual process
of following hunches/intuition and experimentation. Deep learning is such a quickly
evolving field that theoretical explanations for the success of a method almost always
follow empirical evidence rather than vice versa.

Neural Architecture Search is a growing subfield of deep learning, attempting
to develop structured and efficient searches for the most optimal neural network
architectures. Although earlier work in Neural Architecture Search (early 2010s and
before) used primarily Bayesian-based methods, modern NAS work involves the usage
of deep learning structures to optimize deep learning architectures. That is, a neural
network system is trained to "design" the best neural network architectures.

Modern Neural Architecture Search methods contain three key components: the
search space, the search strategy, and the evaluation strategy (Figure 5-10).

Figure 5-10. *Relationship between three key components of Neural Architecture
Search – the search space, the search strategy, and evaluation strategy*

This is a similar structure to how Bayesian optimization frameworks perform optimization. However, NAS systems are differentiated from general optimization frameworks in that they do not approach neural network architecture optimization as a black-box problem. Neural Architecture Search methods take advantage of domain knowledge about neural network architecture representation and optimization by building it into the design of all three components.

Representing the search space of neural network architectures is an interesting problem. The search space must not only be capable of representing a wide array of neural network architectures but also must be set up in a way that enables the search strategy to navigate the search space to sample new promising architectures – there must be some concept of distance (i.e., certain architectures are "closer" to others, Figure 5-11).

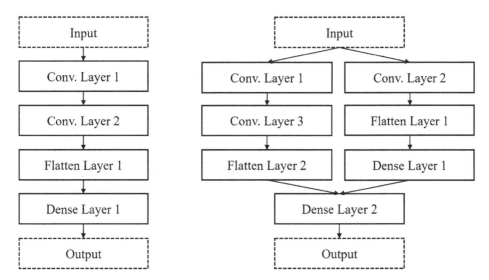

Figure 5-11. *Left – a sequential topology; right – a more complex nonlinear topology*

Moreover, the search space must be capable of representing both linear and nonlinear neural network topologies, which further complicates the organization of such a search space.

Note that Neural Architecture Search systems go to tremendous trouble to represent neural networks in seemingly contrived ways primarily for the purposes of the *search strategy* component, not the search space itself. If the only purpose of the search space component is to represent the model, current graph network implementations more than suffice. However, it's incredibly difficult to create a search strategy that is able to effectively output and work with a neural network architecture in that exact format. The medium of the search space representation allows the search strategy to output a *representation*, which can be used to build and evaluate the corresponding model. Moreover, limiting the search space to only certain architecture designs likely to be successful forces the search strategy to sample from and explore more high-potential architectures.

Perhaps the most simple representation of neural network topologies is to represent them as a sequential string of structured information with many rules to encode and decode the representation (Figure 5-12). If a "block" of information indicates the existence of a convolutional layer, for instance, then other blocks must follow containing information regarding parameters in the convolutional layer, like the kernel size and number of filters.

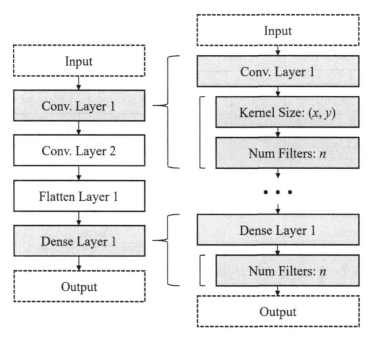

Figure 5-12. *Sequential representation of a linear neural network topology*

Additional adaptations can be made to represent more complex nonlinear topologies and recurrent neural networks by incorporating indices and "anchor points" into the sequential string of information. Skip connections can be modeled via an attention-style mechanism, in which the modeling agent can add skip connections between any two anchor points. The sequential string representation is powerful in that any neural network architecture – regardless of how complex – can be represented in and rebuilt from this format, even if it takes a very long sequence of information.

This sort of sequential representation was used to represent the search spaces of earlier NAS work by Barret Zoph and Quoc V. Le in 2017. The sequential representation of the search space, while being powerful (i.e., it can model many neural network architectures), proved not to be very efficient. Because the search space is so large, however, sequential representations of neural networks proved to be inefficient to navigate.

Cell-based representations of the architecture search space (Figure 5-13) have decreased representative power (i.e., they cannot represent as many architectures as a sequential representation can), but they have proven to be more efficient to navigate. The Neural Architecture Search algorithm learns the architecture of a cell structure, for instance, by selecting operations to fill in a blank "template cell." The cell structure is then repeated in the final neural network architecture. Multiple different types of cell types can be used, and cells can be stacked together in nonlinear ways.

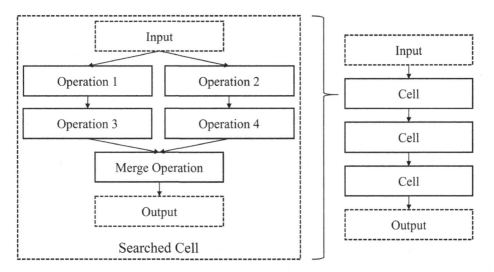

Figure 5-13. *Cell-based neural network architecture representation. See NASNet for an example of a cell-based space*

While the representative power of cell-based search spaces is significantly smaller than that of sequential representations (a network must be built with repeated segments to be efficiently represented cell-wise), it has yielded better performing architectures with shorter resource investment. Moreover, these learned cells can be rearranged, selectively chosen, and transferred to other data contexts and tasks.

The search strategy operates within the search space to find the most optimal representation of the neural network. As discussed prior, the search strategy employed is dependent on the design of the search space representation. For instance, a sequential representation can be modeled with a recurrent based neural network, which takes in previous elements of the network and predicts the next corresponding piece of information (Figure 5-14).

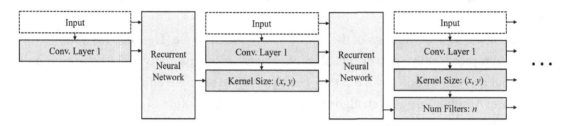

Figure 5-14. *Example recurrent neural network-based search strategy with a sequential information-based search space. This sort of design was used in early 2017 Neural Architecture Search work by Barret Zoph and Quoc. V. Le*

Reinforcement learning is a commonly used search strategy for Neural Architecture Search. In the preceding example, for instance, the recurrent based neural network functions as the *agent* and its parameters function as the *policy*. The agent's policy is iteratively updated to maximize the expected performance of the generated architecture.

Most search strategy methods face a problem: search space representations are discrete, not continuous, but gradient-based search strategies cannot operate on purely discrete problems. Thus, search strategies seeking to take advantage of gradients involve some operation to differentiate discrete operations.

For example, the Differentiable Architecture Search (DARTS) (Figure 5-15) search strategy uses *continuous relaxation*. Every potential operation (e.g., convolution, pooling, activation, etc.) that could be performed to link one "block" to another (think of these not as blocks of layers, but instead as network "anchor points") is represented in a graph. Each operation is associated with a weight. The operation between any two blocks with the highest weight is built into the final model, and a gradient-based method is used to find the weight associations that lead to the selected model with the best performance.

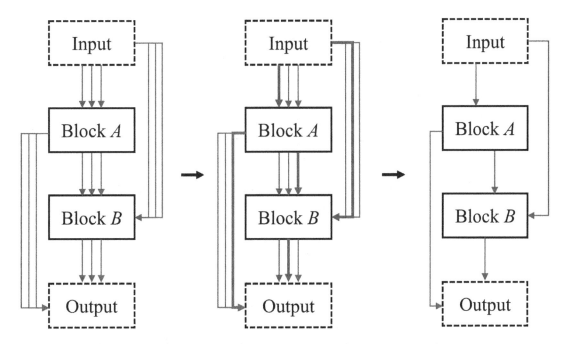

Figure 5-15. *Visualization of continuous relaxation in DARTS. Blue, green, and yellow connections represent potential operations that could be performed on blocks, like a convolution with a 5x5 filter size and 56 filters*

Evolutionary search strategy methods don't require this discrete-to-continuous mapping mechanism, however. Evolutionary algorithms repeatedly mutate, evaluate, and select models to obtain the best performing designs. While evolutionary search was one of the first proposed Neural Architecture Search designs in the early 2000s, genetic-based NAS methods continue to yield promising results in the modern context. Modern evolutionary search strategies generally use more specialized search spaces and mutation strategies to decrease the inefficiency often associated with evolution-based methods.

The simplest method to evaluate the performance of a search strategy is to train the proposed network architecture to completion and to evaluate the performance. This direct method of neural network evaluation suffices and is the most precise method of evaluation, but it is costly on computational and time resources. Unless the searching strategy requires relatively few models to be searched to arrive at a good solution, it's usually infeasible to directly evaluate all proposed architectures.

A proxy evaluation method is a method used to indicate the direct evaluation performance. Several techniques exist:

- *Training on a smaller sampled dataset*: Rather than training the model on the entire dataset, train proposed architectures on a smaller, sampled dataset. The dataset can be selected randomly or to be representative of different "components" of the data (e.g., equal/proportional quantities per label or data cluster).

- *Train a smaller-scaled* version of the architecture: For architecture searching strategies involving predicting a cell-based architecture, the proposed architecture used for evaluation can be scaled down (i.e., fewer number of repeats or less complexity).

- *Predicting test performance curve*: The proposed architecture is trained for a few epochs, and a time series regression model is trained to extrapolate out the expected future performance. The maximum of the extrapolated future performance is taken to be the predicted performance of the proposed architecture.

See different approaches toward the three components of Neural Architecture Search visually mapped in Figure 5-16.

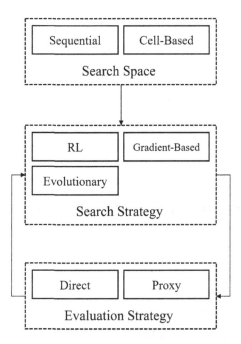

Figure 5-16. *Demonstration of methods discussed for the search space, search strategy, and evaluation strategy components of NAS. Note that the discussed methods, of course, only cover some of modern NAS research*

For more information on advances in NAS, see this chapter's case studies, which detail three key advances in Neural Architecture Search.

Neural Architecture Search is currently a quite cutting-edge area of deep learning research. Many of the NAS methods discussed still require massive quantities of time and computational resources and are not suitable for general usage in the same way architectures like convolutional neural networks are. We will use Auto-Keras – a library that efficiently adapts the Sequential Model-Based Optimization framework for architecture optimization – for NAS.

Auto-Keras

There are many auto-ML libraries like PyCaret, H2O, and Azure; for the purposes of this book, we use Auto-Keras, an auto-ML library built natively upon Keras. Auto-Keras demonstrates the progressive disclosure of complexity principle, meaning that users can both build incredibly simple searches and run more complex operations.

To install, use `pip install auto-keras`.

Auto-Keras System

The Auto-Keras Neural Architecture Search system is one of very few easily accessible NAS libraries, as of the writing of this book – even state-of-the-art NAS designs are still too computationally expensive and intricate to be feasibly written into an approachable package.

Auto-Keras uses Sequential Model-Based Optimization (SMBO), which was earlier presented as a formal framework for understanding Bayesian optimization.[1] However, while generalized SMBO is designed to solve black-box problems, like the Tree-structured Parzen Estimator strategy used in Hyperopt, Auto-Keras exploits domain knowledge about the problem domain – neural network architectures – to develop more efficient SMBO components.

While the Tree-structured Parzen Estimator strategy employs the TPE surrogate model, Auto-Keras utilizes another commonly used model, the Gaussian process. Like all surrogate models used in SMBO, the Gaussian process probabilistically represents knowledge about the true function. Unlike TPE, however, the Gaussian process does so by learning the probability distribution of functions that could feasibly represent the sampled data (Figure 5-17). A function that fits the sampled data well will be associated with a high probability, whereas a function that fits the data poorly is associated with a low probability. The mean of this probability distribution is the most representative function. (Note that there is an infinite quantity of existing functions, so few functions have significant probabilities.)

---

[1] Haifeng Jin, Qingquan Song, and Xia Hu, "Auto-Keras: An Efficient Neural Architecture Search System," 2019.

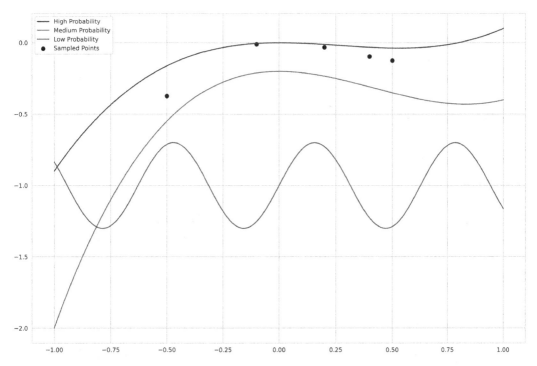

Figure 5-17. *Simplified representation of the idea behind Gaussian processes –*
different fitted functions on sampled points and their associated probabilities.
In practice, you probably won't see a polynomial fit next to a sinusoidal fit in the
same Gaussian process operation, but it is included for the sake of conceptual
visualization

Gaussian processes require a probability distribution of functions in Euclidean
space, but it is very difficult to map neural network architectures to vectors in the
Euclidean space. To address this problem, Auto-Keras uses the *edit distance neural*
network kernel (Figure 5-18), which quantifies the minimum number of edits required to
morph some network f_a into another network f_b.

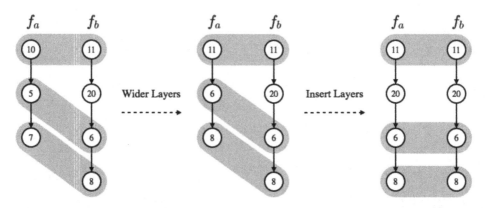

Figure 5-18. *Figure showing morphing process used in the edit distance neural network kernel. First, the second and third layers in f_a are widened to contain the same number of neurons as the corresponding layers in f_b. Then, another layer is added to f_a. Created by Auto-Keras authors*

The edit distance neural network kernel allows for a quantification of similarity between neural network structures, which – roughly speaking – provides the necessary link from the discrete neural network architecture search space to the Euclidean space of the Gaussian process. Using the edit distance kernel and a correspondingly designed acquisition function, the Auto-Keras NAS algorithm is able to control the crucial exploit/explore dynamic – it samples network architectures with a low edit distance with respect to successful networks to exploit and samples network architectures with a high edit distance with respect to poorer performing networks to explore.

Auto-Keras trains each model to completion for some number of user-specified epochs (i.e., it uses a direct evaluation method rather than a proxy method), but its Bayesian character requires a fewer number of networks to be sampled and trained, provided the search space is well defined. In experiments carried out by Auto-Keras' authors, Auto-Keras achieves a lower classification error than state-of-the-art network morphism and Bayesian-based approaches to Neural Architecture Search in the benchmark MNIST, CIFAR-10, and Fashion datasets.

The Auto-Keras API further employs the usage of parallelism between CPU and GPU; the GPU is used to train the generated model architectures, whereas the CPU is used to perform searching and updating (Figure 5-19). Moreover, Auto-Keras utilizes a memory estimation function to enable efficient GPU usage and to prevent GPU memory crashes (Figure 5-20).

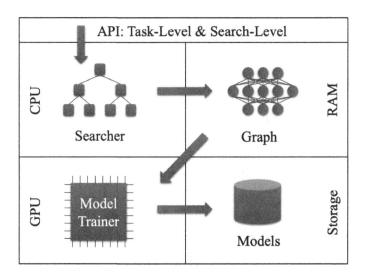

Figure 5-19. *Auto-Keras' API task-level and search-level operations in hardware. Created by Auto-Keras authors*

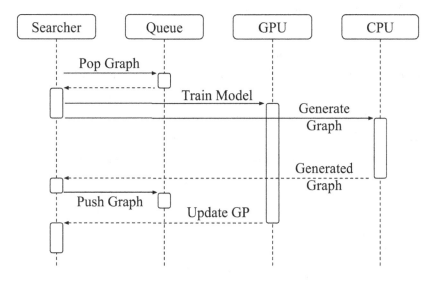

Figure 5-20. *Auto-Keras' system design – relationship between searcher, queue, GPU, and CPU usage. Proposed graphs are generated in CPU, kept in the queue, and trained in GPU*

Both in its design and implementation, Auto-Keras is a key library in making Neural Architecture Search algorithms more efficient and accessible.

Simple NAS

The simplest form of Neural Architecture Search is to define the inputs and outputs and allow the NAS algorithm to automatically determine all processing layers to "connect" input to output. Let's consider building a neural network for an image classification task. Auto-Keras follows a very similar syntax to the Keras Functional API. For this particular task, we need to define three key elements, linked together via function notation:

1. *The input node*: In this particular case, the input is an image input, so we use ak.ImageInput(), which accepts numpy arrays and TensorFlow datasets. For other input data types, use ak.StructuredDataInput() for tabular data (accepts pandas DataFrames in addition to numpy arrays and TensorFlow datasets), ak.TextInput() for text data (must be a numpy array or TensorFlow dataset of strings; Auto-Keras performs vectorization automatically), or ak.Input() as a general input method accepting tensor data from numpy arrays or TensorFlow datasets from all contexts. Using this last method comes at the cost of helpful preprocessing and linkage that Auto-Keras performs automatically when a context-specific data input is specified. These are known in Auto-Keras terminology as *node* objects. There is no need to specify the input shape; Auto-Keras automatically infers it from the passed data and constructs architectures that are valid with respect to the input data shape.

2. *The processing block*: You can think of blocks in Auto-Keras as supercharged clumps of Keras layers – they perform similar functions as groups of layers, but are integrated into the Auto-Keras NAS framework such that key parameters (e.g., number of layers in the block, which layers are in the block, parameters for each layer) can be left unspecified and are automatically tuned. For instance, the ak.ConvBlock() block consists of the standard "vanilla" series of convolutional, max pooling, dropout, and activation layers – the specific number, sequence, and types of layers can be left to the NAS algorithm. Other blocks, like ak.ResNetBlock(), generate a ResNet model; the NAS algorithm tunes factors like which version of the ResNet model to use

and whether to enable pretrained ImageNet weights or not.

* Blocks represent the primary components of neural networks; Auto-Keras' design is centered around manipulating blocks, which allows for more high-level manipulation of neural network structure. If you want to be even more general, you can use context-specific blocks like `ak.ImageBlock()`, which will automatically choose which image-based block to use (e.g., vanilla convolutional block, ResNet block, etc.).

3. *The head/output block*: Whereas the input node defines what Auto-Keras should expect to be passed into the input of the architecture, the head block defines what type of prediction task the architecture should be participating in by determining two key factors: the activation of the output layer and the loss function. For instance, in a classification task, `ak.ClassificationHead()` is used; this block automatically infers the nature of the classification head (binary or multiclass classification) and correspondingly imposes limits on the architecture (sigmoid and binary cross-entropy for binary classification vs. softmax and categorical cross-entropy for multiclass classification). If it detects "raw" labels (i.e., labels that have not been preprocessed), Auto-Keras will automatically perform binary encoding, one-hot encoding, or any other encoding procedure required to conform the data to the inferred prediction task. It's usually best to make sure Auto-Keras does not need to make any drastic changes based on its inferences on your intent, however, as a precaution. Similarly, use `ak.RegressionHead()` for regression problems.

To build an incredibly simple and general image classification model, we can begin by importing Auto-Keras and defining the three key components of the neural network architecture in functional relation to one another (Listing 5-18).

Listing 5-18. Simple input-block-head Auto-Keras architecture

```
import autokeras as ak
inp = ak.ImageInput()
imageblock = ak.ImageBlock()(inp)
output = ak.ClassificationHead()(imageblock)
```

Note   For text data, use ak.TextBlock(), which chooses from vanilla, transformer, or n-gram text-processing blocks. Auto-Keras will automatically choose a vectorizer based on the processing block used. For tabular/structured data, use ak.StructuredDataBlock(); Auto-Keras will automatically perform categorical encoding and normalization. This will need to be followed by a processing block like ak.DenseBlock(), which stacks FC layers together.

Just as in the Keras Functional API, these layers can be aggregated into a "model" by specifying the input and output layers (Listing 5-19). The max_trials parameter indicates the maximum number of different Keras models to try, although the search may conclude before reaching that quantity. The "model" can then be fitted on the data; the epochs parameter represents the number of epochs to train each candidate Keras model on.

Listing 5-19. Aggregating defined components into an Auto-Model and fitting

```
search = ak.AutoModel(
    inputs=inp, outputs=output, max_trials=30
)
search.fit(x_train, y_train, epochs=10)
```

During the search, Auto-Keras will not only automatically determine which type of image-based block to choose but also various normalization and augmentation methods to optimize model performance.

You may also see some error messages printed when working with Auto-Keras for debugging purposes – as long as the code keeps on running, it's generally safe to ignore the warnings.

It's important to realize that this "model" is not really a model, but a searching object that acts as a template upon which various model candidates are created and evaluated. In order to extract the best model after training, call best_model = search.export_ model(). You can then use best_model.summary() or plot_model(best_model) to list or visualize the architecture of the best model from the search (Figure 5-21).

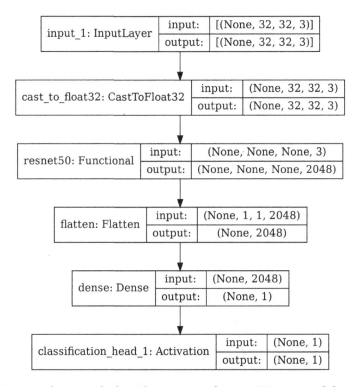

Figure 5-21. *Example sampled architecture of Auto-Keras model search*

Unfortunately, such an approach – simply defining the inputs and outputs and letting Neural Architecture Search figure out the rest – is unlikely to yield good results in a feasible amount of time. Recall that every parameter you leave untuned is another parameter the NAS algorithm must consider in its optimization, which expands the number of trials it needs.

NAS with Custom Search Space

We can design a more practical architecture in terms of time to reach a good solution and computational burden by providing certain limitations on the search space using strategies that we know work.

Recall that using Auto-Keras means playing with higher-level blocks or components. We know the following image recognition pipeline to be successful (see Chapter 2 on transfer learning and pretrained models, Figure 5-22).

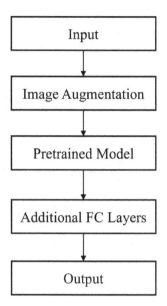

Figure 5-22. *General component-level design of image recognition models*

We can build these blocks using Auto-Keras blocks:

1. ak.ImageInput(), as discussed prior, is the input node.

2. ak.ImageAugmentation() is an augmentation block that performs
    various image augmentation procedures, like random flipping,
    zooming, and rotating. Augmentation parameters, like whether
    to randomly perform horizontal flips or the range with which to
    select a factor to zoom or rotate, are automatically tuned by
    Auto-Keras if left unspecified.

3. ak.ResNetBlock(), as discussed prior, is a ResNet architecture
    with only two parameters – which version of ResNet to use
    (v1 or v2) and whether to initialize with ImageNet pretrained
    weights or not. This serves as the pretrained model component of
    our image model design.

4. ak.DenseBlock() is a block consisting of fully connected layers.
    If left unspecified, Auto-Keras tunes four parameters: the number
    of layers, whether or not to use batch normalization in between
    layers, the number of units in each layer, and the dropout rate to
    use (a rate of 0 indicates not to use dropout).

5. ak.ClassificationHead(), as discussed prior, is the head block specifying the loss function to use and the activation of the last output layer.

Let's define these functionally in relation to one another and specify parameters which we have a good idea of what we want (Listing 5-20). For instance, in augmentation, we may know that we don't want to flip vertically or horizontally (e.g., in the MNIST dataset, flipping some digits will necessitate changing their label). We also know that we want a translation factor of 10% the image width – not too large nor small. However, we're not quite sure which zoom or contrast factor is best; these parameters can be left as None and will be automatically tuned by Auto-Keras. Similarly, we would like the ResNet block to use pretrained weights – enough to further process the output of the ResNet block, but not enough to cause network length and overfitting problems.

Listing 5-20. Defining an architecture with more complex custom search space, with all parameters specified

```
inp = ak.ImageInput()
aug = ak.ImageAugmentation(translation_factor=0.1,
                           vertical_flip=False,
                           horizontal_flip=False,
                           rotation_factor=None,
                           zoom_factor=None,
                           contrast_factor=None)(inp)
resnetblock = ak.ResNetBlock(pretrained=True,
                             version=None)(aug)
denseblock = ak.DenseBlock(num_layers=None,
                           use_bn=None,
                           num_units=None,
                           dropout=None)(resnetblock)
output = ak.ClassificationHead()(xceptionblock)
```

Since Auto-Keras leaves all parameters at None by default, we can more compactly represent the same specified parameters by removing all parameters we set explicitly to None (Listing 5-21).

Listing 5-21. Defining an architecture with more complex custom search space, with only relevant parameters specified

```
inp = ak.ImageInput()
aug = ak.ImageAugmentation(translation_factor=0.1,
                           vertical_flip=False,
                           horizontal_flip=True)(inp)
resnetblock = ak.ResNetBlock(pretrained=True,
                             version=None)(aug)
denseblock = ak.DenseBlock()(resnetblock)
output = ak.ClassificationHead()(xceptionblock)
```

The layers can then be aggregated into an ak.AutoModel (visualized in Figure 5-23) and fitted accordingly. By specifying a custom search space, you dramatically increase the chance that you will arrive at a satisfactory solution in less chance.

Note   For text data, use ak.TextBlock(), use ak.Embedding() for embedding (can use GloVe, fastText, or word2vec pretrained embeddings as pretraining), and use ak.RNNBlock() for recurrent layers. For tabular/structured data, use ak.CategoricalToNumerical() for numerical encoding of categorical features in addition to standard processing blocks, like ak.DenseBlock().

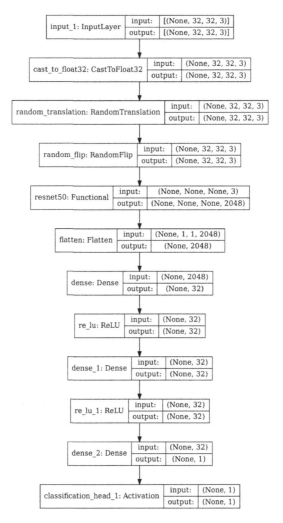

Figure 5-23. *Example sampled architecture of Auto-Keras model search*

NAS with Nonlinear Topology

Because Auto-Keras is built using Functional API-like syntax, we can also define broadly nonlinear topologies using components. For instance, rather than passing the input through only one pretrained model block, we could pass it through two pretrained models to obtain their "insights"/"perspectives" on the inputs and then merge and process the two together (Figure 5-24).

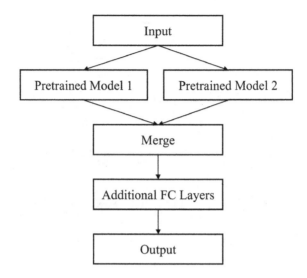

Figure 5-24. *Component-wise plan for implementing a nonlinear topology in Auto-Keras*

We can express this idea using Function API-like syntax, aggregating and fitting afterward (Listing 5-22, Figure 5-25).

Listing 5-22. Building component-wise topologically nonlinear Auto-Keras designs

```
inp = ak.ImageInput()
resnetblock = ak.ResNetBlock(pretrained=True)(inp)
xceptionblock = ak.XceptionBlock(pretrained=True)(inp)
merge = ak.Merge()([resnetblock, xceptionblock])
denseblock = ak.DenseBlock()(merge)
output = ak.ClassificationHead()(denseblock)
```

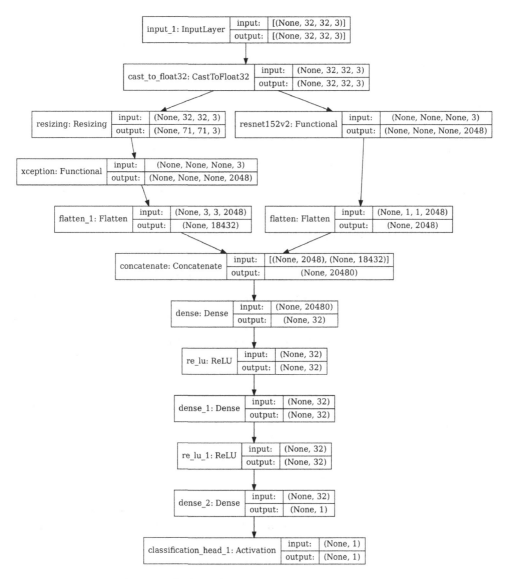

Figure 5-25. *Example sampled architecture of Auto-Keras model search*

Case Studies

These three case studies discuss three different approaches to developing more successful and efficient Neural Architecture Search systems, building upon topics discussed both in the NAS and the Bayesian optimization sections of this chapter.

As pillars of the rapidly developing Neural Architecture Search research frontier, these case studies will themselves serve as the foundations upon future work in the automation of more powerful neural network architectures.

NASNet

The NASNet search space, proposed by Barret Zoph, Vijay Vasudevan, Jonathon Shlens, and Quoc V. Le,[2] is cell-based. The NAS algorithm learns two types of cells: normal and reduction cells. Normal cells make no change to the shape of the feature map (i.e., input and output feature maps have identical shapes), whereas reduction cells halve the width and height of the input feature map.

Cells are built as complex topological combinations of blocks, which are small, architecturally predefined "templates" that contain several "blank operation slots" that are learned by the NAS algorithm and are arranged together into a network cell. A cell is defined as a certain fixed number of these blocks B. ($B=5$ in the authors' experiments.) These cells can then be sequentially stacked to form a neural network (Figure 5-26).

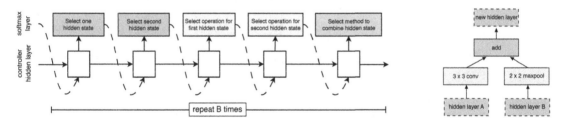

Figure 5-26. *The generation of architectures via a recurrent model, which outputs hidden states, operations, and a merging method to select in the design of a cell*

A recurrent neural network is used to iteratively generate these blocks by choosing two hidden states to combine in the operation, two operations to apply to the two hidden states individually, and one merging method. Operations include the identity operation, standard square kernel convolutions, rectangular kernel convolutions, dilated convolutions (the kernel is widened or "inflated" with empty spaces in between kernel

---

[2] Barret Zoph, Vijay Vasudevan, Jonathon Shlens, and Quoc V. Le, "Learning Transferable Architectures for Scalable Image Recognition," 2018. Paper link: `https://arxiv.org/pdf/1707.07012.pdf`.

weights), separable convolutions (convolutions that apply not only to spatial but also depth dimensions), pooling, and more. Two branches can be merged either through adding or concatenation. The network is trained via reinforcement learning methods to maximize the test performance of the resulting neural network architecture proposal (Figure 5-27).

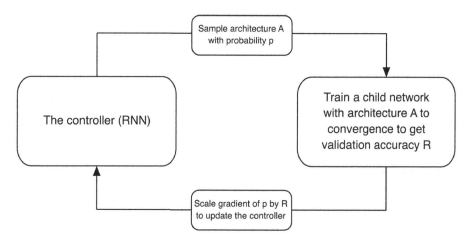

Figure 5-27. *Relationship between the recurrent based controller and the proposed architecture (the child network). The controller is updated to maximize the validation performance of the child network. Created by NASNet authors*

Moreover, because the recurrent neural network is able to select which two outputs of previously constructed cell outputs to perform the cell operation on, it is able to construct incredibly elaborate architectures in an elegant, recursive manner (Figures 5-28 and 5-29).

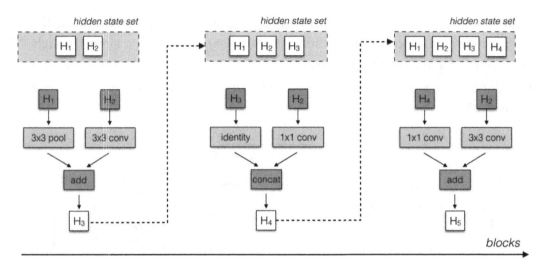

Figure 5-28. *Example selection of hidden states and operations via recurrent style generation. Created by NASNet authors*

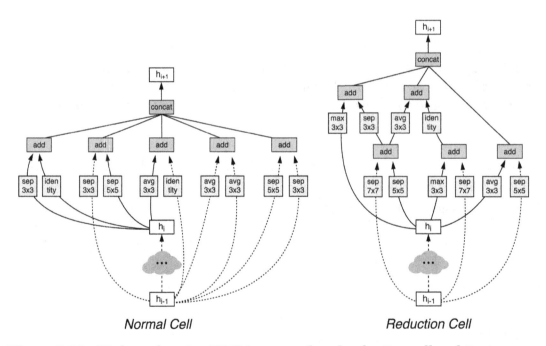

Figure 5-29. *High-performing NASNet normal and reduction cell architectures on ImageNet. Created by NASNet authors*

The derived normal and reduction cells can be stacked together in different lengths to suit different datasets (Figure 5-30). A network built for the CIFAR-10 dataset, for instance, with its incredibly small 32x32 resolution, uses fewer reduction cells than a network built for the ImageNet architecture. This sort of ease in transferring the results of an intensive Neural Architecture Search to all sorts of different contexts greatly increases its practicality.

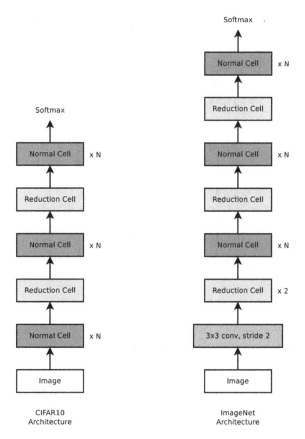

Figure 5-30. *Stacking normal and reduction cells to suit a certain dataset image size. Created by NASNet authors*

The NASNet architecture family is constructed from the best performing normal and reduction cells. It manages to obtain better top 1 and top 5 accuracy in ImageNet classification against previous architectural giants like Inception and more recently proposed architectures with a fewer number of parameters (Table 5-1).

Table 5-1. *Performance of various sizes of NASNet (sizes determined by stacking different combinations and lengths of normal and reduction cells) against similarly sized models*

Model	# Parameters	Performance	
		Top 1 Acc.	Top 5 Acc.
Small-sized models			
– InceptionV2	11.2 M	74.8%	92.2%
– *Small NASNet*	*10.9 M*	*78.6%*	*94.2%*
Medium-sized models			
– InceptionV3	23.8 M	78.8%	94.%
– Xception	22.8 M	79.0%	94.5%
– Inception ResNetV2	55.8 M	80.1%	95.1%
– *Medium NASNet*	*22.6 M*	*80.8%*	*95.3%*
Large-sized models			
– ResNeXt	83.6 M	80.9%	95.6%
– PolyNet	92 M	81.3%	95.8%
– DPN	79.5 M	81.5%	95.8%
– *Large NASNet*	*88.9 M*	*82.7%*	*96.2%*

The NASNet architecture (not the search process) with ImageNet weights is available in Keras. It comes in two versions, NASNet Large and NASNet Mobile, which are scaled versions of the best performing learned cell architectures from the NASNet search space. The architectures are available in `keras.applications` at

- `keras.applications.nasnet.NASNetMobile`: 23 MB storage size with 5.3 m parameters.

- `keras.applications.nasnet.NASNetLarge`: 343 MB storage size with 88.9 m parameters. (NASNet Large, as of the writing of this book, holds the highest ImageNet top 1 and top 5 accuracy of all `keras.applications` models with such reported metrics.)

See Chapter 2 on usage of pretrained models.

Even given NASNet's advances in developing high-performing cell architectures, such results required hundreds of GPUs and 3–4 days of training in Google's powerful laboratories. Other advances worked to build more computationally accessible search operations.

Progressive Neural Architecture Search

Chenxi Liu, along with other coauthors at Johns Hopkins University, Google AI, and Stanford University, proposes the *Progressive Neural Architecture Search* (PNAS).[3] True to its naming, PNAS adopts a progressive approach to the building of neural network architectures from simple to complex. Moreover, PNAS interestingly combines many of the earlier discussed Neural Architecture Search methods into one cohesive, efficient approach: a cell-based search space, a Sequential Model-Based Optimization search strategy, and proxy evaluation.

The search space used by Progressive Neural Architecture Search is very similar to that of the NASNet design, with one key difference: rather than learning two different cells (a normal and reduction cell), PNAS only learns one type of "normal cell" (Figure 5-31). A "reduction cell" is formed by using a normal cell with a stride of 2. This slightly reduces the size of the search space relative to that of NASNet.

---

[3] Chenxi Liu, Barret Zoph, Maxim Neumann, Jonathon Shlens, Wei Hua, Li-Jia Li, Li Fei-Fei, Alan Yuille, Jonathan Huang, and Kevin Murphy, "Progressive Neural Architecture Search," 2018. Paper link: `https://arxiv.org/pdf/1712.00559.pdf`. Code link: `https://github.com/titu1994/progressive-neural-architecture-search` (Keras/TensorFlow implementation).

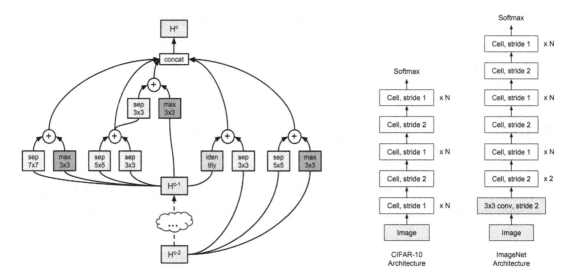

Figure 5-31. *Progressive Neural Architecture Search cell design. Left – high-performing PNAS cell design. Right – examples of how PNAS cell architectures can be stacked with different stride lengths to adapt to datasets of different sizes. Created by PNAS authors*

PNAS makes use of a Sequential Model-Based Optimization search strategy, in which the most "promising" model proposals are selected for evaluation, in conjunction with a proxy evaluator.

The proxy evaluator, an LSTM, is trained to read an information sequence representing the architecture of a proposed model and to predict the performance of the proposed model. A recurrent based model was chosen for its ability to handle variable length inputs. Note that the proxy evaluator is trained on a very small dataset (the label – the performance of a proposed model – is expensive to obtain), so an ensemble of LSTMs trained on a subset of the data is used to support generalization and decrease variation. An RNN-based method's predicted performance of candidate model architectures can reach as high as 0.996 Spearman rank correlation with the true performance rank.

Progressive Neural Architecture Search (Figure 5-32) begins with the simplest cell architectures first, which consist of only one block. Then, each cell is expanded by adding another block to the cell architecture. As the number of blocks in the cell architecture increases, the number of candidates and the resources required to train grows exponentially, so these candidate models cannot all be trained. This is where the proxy evaluator comes in – the proxy evaluator evaluates hundreds of thousands of proposed architectures in negligible time, and the top *K* most promising architectures (highest performing according to the proxy evaluator) are sampled for direct evaluation. The results of these architectures are in turn used as additional training data to update the proxy evaluator. This is repeated until a satisfactory number of blocks per cell is reached.

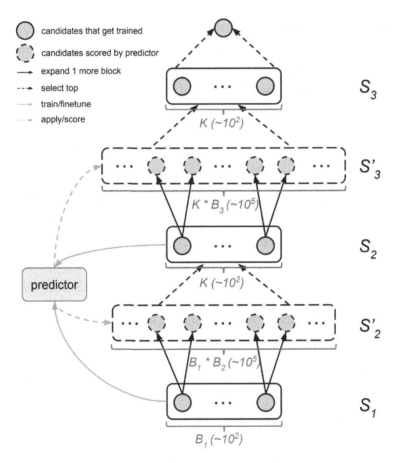

Figure 5-32. *Visualization of the PNAS search and evaluation process. From S_1 to S'_2 trained cell architectures are expanded by one block. These generated cell architectures are evaluated via a proxy evaluator (labeled "predictor" in the visualization) and the top few are selected for training in S_2. Created by PNAS authors. The performance of the generated architectures is used to update the proxy evaluator. The process repeats*

The proxy evaluator functions as the surrogate function in Sequential Model-Based Optimization, used to perform sampling and updated using sampled results to be more accurate. Its progressive design allows for computational efficiency – if a smaller number of blocks per cell returns good performance, we have saved ourselves from needing to run through architectures with higher blocks per cell; even if a smaller number of blocks per cell does not yield good results, it functions as live training for the proxy evaluator.

This design allows Progressive Neural Architecture Search to yield significant speedups over previous NAS methods, reducing the number of models that need to be trained by the thousands while reaching the same accuracy. Moreover, its cell-based design, like NASNet, allows for transferability of the cell design across different datasets (Table 5-2).

Table 5-2. *Performance of PNAS against earlier work by Barret and Zoph in 2017, in which reinforcement learning is used to optimize a RNN to generate sequential representations of CNN architectures. Note that this is different from the closely related work by Zoph, Vasudevan, Shlens, and Le in 2018 on NASNet, which uses a cell-based representation. "# models trained on <method>" indicates the number of models the method trains to reach the listed corresponding performance. PNAS can reach almost a fivefold decrease in the number of models trained*

Cells per Block	Top	Accuracy	# Models Trained by PNAS	# Models Trained by NAS
5	1	0.9183	1160	5808
5	5	0.9161	1160	4100
5	25	0.9136	1160	3654

Efficient Neural Architecture Search

While proxy evaluation in Progressive Neural Architecture Search allows for quick prediction on the potential of a proposed model architecture and thus decreases the number of models that needs to be trained for good performance, the computational and time bottleneck in the process still remains in the training stage. Hieu Pham and Melody Y. Guan, along with Barret Zoph, Quoc V. Le, and Jeff Dean, put forth the *Efficient*

Neural Architecture Search (ENAS)[4] method, which attempts to decrease the time needed to obtain accurate measurements on the performance of a candidate model by forcing *weight sharing* across all candidate architectures during training.

ENAS uses a similar reinforcement learning and recurrent based architecture generation model as the NASNet creators use, with one key difference: rather than predefining the "template" or "slots" of the cell and training the architecture generation model to identify which operations to "fill in" the "slots" with, in ENAS the controller model not only identifies which operations to choose but also how operations are connected.

Candidate network architectures can be represented as directed acyclic graphs (DAGs) or a graph with directions but no cycles (i.e., you cannot end up in a loop by following directed connections between nodes). A DAG with *N* nodes is initialized (Figure 5-33, *N*=4) in which each node is connected to every other node. This "fully connected" DAG represents the search space for that block, and architectures can be sampled by selecting *sub-graphs* within the full DAG. Nodes that are included in the sampled sub-graphs are attached to an operation, and all the "dead ends" of the graph are averaged and considered outputs of the cell.

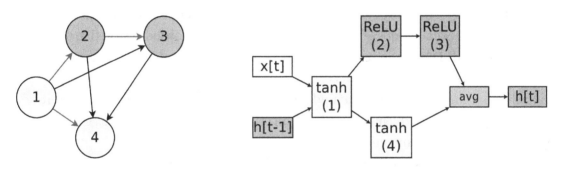

Figure 5-33. *Left: "Fully connected" DAG with selected sub-graph shown in red. Right: example selected architecture based on sampled sub-graph*

These architectures are generated by an LSTM trained using reinforcement learning methods to maximize the validation performance of generated architectures (Figure 5-34). The LSTM selects sub-graphs by outputting the index of the node

[4] Hieu Pham, Melody Y. Guan, Barret Zoph, Quoc V. Le, and Jeff Dean, "Efficient Neural Architecture Search via Parameter Sharing," 2018. Paper link: `https://arxiv.org/pdf/1802.03268.pdf`. Code link: `https://github.com/melodyguan/enas` (TensorFlow implementation).

that the node being generated currently will be attached to. This sort of generation procedure can be applied both to generate cells that will be stacked into full architectures and to produce an entire architecture single handedly.

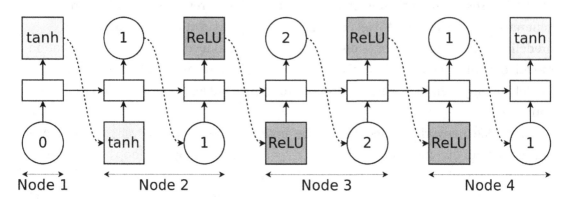

Figure 5-34. *Recurrent model selection of sub-graphs from the "fully connected" DAG for the example model shown in Figure 5-30*

The conceptual understanding of all sampled architectures as sub-graphs of the "super-graph," "fully connected" DAG is crucial as the underlying basis of ENAS' usage of weight sharing. The "fully connected" DAG represents a series of knowledge-based relationships between nodes; for maximum efficiency, the knowledge stored in the "fully connected" DAG's connections should be transferred to selected sub-graphs. Thus, all proposed sub-graphs that contain the same connection will share the same value for that connection. Gradient updates performed on one proposed architecture's connections are also performed identically on other proposed architectures with the same corresponding connections.

By sharing weights, proposed models can "learn from one another" by updating weights they have in common via the insights derived by another model architecture. Moreover, it serves as a rough approximation as to how models with similar architectures would have developed regardless under the same conditions.

This aggressive weight sharing "approximation" allows for massive quantities of proposed model architectures to be trained with much smaller computation and time consumption. Once the child models are trained, they are evaluated on a small batch of validation data and the most promising child model is trained from scratch.

With additional regularization, Efficient Neural Architecture Search allows for a tremendous speedup from several days to a fraction of a day with very few GPUs (Table 5-3). ENAS is a huge step toward making Neural Architecture Search a reality outside of high-computation laboratories.

Table 5-3. *Performance of ENAS against the results of other Neural Architecture Search methods. CutOut is an image augmentation method for regularization in which square regions of the input are randomly masked during training. CutOut is applied to NASNet and ENAS to increase performance of final architecture*

Method	GPUs	Time (Days)	Params	Error
Hierarchical NAS	200	1.5	61.3 m	3.63%
Micro NAS with Q-Learning	32	3	–	3.60%
Progressive NAS	100	1.5	3.2 m	3.63%
NASNet-A	450	3–4	3.3 m	3.41%
NASNet-A + CutOut	450	3–4	3.3 m	2.65%
ENAS	1	0.45	4.6 m	3.54%
ENAS + CutOut	1	0.45	4.6 m	2.89%

Key Points

In this chapter, we discussed the intuition, theory, and implementation for general hyperparameter optimization and Neural Architecture Search and their implementations in Hyperopt, Hyperas, and Auto-Keras:

- In meta-optimization, a controller model optimizes the structural parameters of a controlled model to maximize the controlled model's performance. It allows for a more structured search for the best "type" of model to train. Meta-optimization methods repeatedly select structural parameters for a proposed controlled model and evaluate their performance. Meta-optimization algorithms used in practice incorporate information about the performance of previously selected structural parameters to inform how the next set of parameters is selected, unlike naïve methods like grid or random search.

- A key balance in meta-optimization is that of the size of the search space. Defining search space to be larger than it needs to significantly expands the computational resources and time required to obtain a good solution, whereas defining too narrow a search space is likely to yield results no different from user-specified parameters (i.e., meta-optimization is not necessary). Be as conservative in determining parameters to be optimized by meta-optimization (i.e., do not be overly redundant in your search space), but ensure that parameters allocated for meta-optimization are "wide" enough to yield significant results.

- Bayesian optimization is a meta-optimization method to address black-box problems in which the only information provided about the objective function is the corresponding output of an input (a "query") and in which queries to the function are expensive to obtain. Bayesian optimization makes use of a surrogate function, which is a probabilistic representation of the objective function. The surrogate function determines how new inputs to the objective function are sampled. The results of these samples in turn affect how the surrogate function is updated. Over time, the surrogate function develops accurate representations of the objective function, from which the optimal set of parameters can be easily derived.

  - Sequential Model-Based Optimization (SMBO) is a formalization of Bayesian optimization and acts as a central component or template against which various model optimization strategies can be formulated and compared.

  - The Tree-structured Parzen Estimator (TPE) strategy is used by Hyperopt and represents the surrogate function via Bayes' rule and a two-distribution threshold-based design. TPE samples from locations with lower objective function outputs.

  - Hyperopt usage consists of three key components: the objective function, the search space, and the search operation. In the context of meta-optimization, the model is built inside the objective function with the sampled parameters. The search space is defined via a dictionary containing hyperopt.hp distributions (normal, log-normal, quantized normal, choice, etc.). Hyperopt can be used to optimize the training procedure, as well as to make fine-tuned optimizations to the model architecture. Hyperas is a Hyperopt wrapper that makes using Hyperopt to optimize various components of neural network design more convenient by removing the need to define a separate search space and independent labels for each parameter.

- Neural Architecture Search (NAS) is the process of automating the engineering of neural network architectures. NAS consists of three key components: the search space, the search strategy, and the evaluation strategy. The search space of a neural network can be represented most simply as a sequential sequence of operations, but it's not as efficient as a cell-based design. Search strategies include reinforcement learning methods (a controller model is trained to find the optimal policy – the parameters of the controlled model) and evolutionary designs. Methods like DARTS map the discrete search space of the neural network into a continuous, differentiable one. Evaluation of sampled parameters can take the form of direct evaluation or via proxy evaluation, in which the performance of the parameters is estimated with fewer resources at the cost of precision.

  - The Auto-Keras system uses Sequential Model-Based Optimization, with a Gaussian process-based surrogate model design and the edit distance neural network kernel to quantify similarity between network structures. Moreover, Auto-Keras is built with GPU-CPU parallelism for optimal efficiency. In terms of user usage, Auto-Keras employs the progressive disclosure of complexity principle, allowing users to build both incredibly simple and more complex architectures with few lines of code. Moreover, because it follows Functional API-like syntax, users can build searchable architectures with nonlinear topologies.

In the next chapter, we will discuss patterns and concepts in the design of successful neural network architectures.

## CHAPTER 6

Successful Neural Network Architecture Design

Design is as much a matter of finding problems as it is solving them.

—Bryan Lawson, Author and Architect

The previous chapter, on meta-optimization, discussed the automation of neural network design, including automated design of neural network architectures. It may seem odd to follow such a chapter on the automation of neural network architectures with a chapter on successful (implicitly, manual) neural network architecture design, but the truth is that deep learning will not reach a point anytime soon (or at all) where the state of design could possibly dismiss the need to possess an understanding of key principles and concepts in neural network architecture design. You've seen, firstly, that Neural Architecture Search – despite making rapid advancements – is still limited in many ways by computational accessibility and reach in the problem domain. Moreover, Neural Architecture Search algorithms themselves need to make implicit architectural design decisions to make the search space more feasible to search, which human NAS designers need to understand and encode. The design of architectures simply cannot be wholly automated away.

© Andre Ye 2022
A. Ye, *Modern Deep Learning Design and Application Development*,
https://doi.org/10.1007/978-1-4842-7413-2_6

The success of transfer learning is earlier and stronger evidence for the proposition that the design of neural network architectures is a study that need not be studied by the average deep learning engineer. After all, if the pretrained model libraries built into TensorFlow and PyTorch don't satisfy the needs of your problem, platforms like the TensorFlow model zoo, GitHub, and pypi host a massive quantity of easily accessible models that can be transferred, slightly adapted with minimal architectural modifications to fit the data shapes and objectives of your problem, and fine-tuned on your particular dataset.

This is partially true – the availability of openly shared model architectures and weights has brought about a decreased need to design large architectures from scratch. However, in practice, it's more likely than not that model architectures you select don't completely align with the context of your problem. Unless an architecture was crafted particularly for your problem domain (and even if it is), usually there is a need for more significant architectural modifications (Figure 6-1).

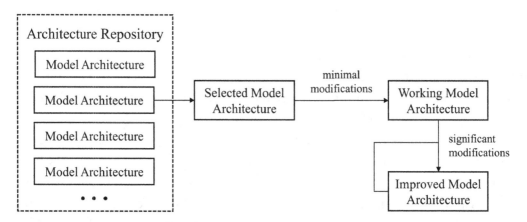

Figure 6-1. *Architectures often still need significant modifications to adapt to your particular problem domain*

The process of making these more significant modifications to adapt the architecture toward success in your problem domain is a continual one; when designing your network, you should repeatedly experiment with the architecture and make architectural adaptations in response to the feedback signals of previous experiments and problems that emerge.

Both constructing successful modifications and successfully integrating them into the general model architecture require a knowledge of successful architecture-building patterns and techniques. This chapter will discuss three key ideas in neural network architecture construction – nonlinear and parallel representation, cell-based design, and network scaling. With knowledge of these concepts, you will not only be able to successfully modify general architectures but also to analytically deconstruct and understand architectures and to construct successful model architectures from scratch – a tool whose value extends not only into the construction of network architectures but cutting-edge fields like NAS.

In order to accomplish these goals, we're going to need to use Keras in a more complex fashion. As the neural network architectures we seek to implement become more intricate, there is an increasing ambiguity of implementation – that is, there are many "correct answers." In order to navigate this ambiguity of implementation more efficiently, we will begin to *compartmentalize*, *automate*, and *parametrize* the defining of network architectures:

- *Compartmentalization*: This concept was introduced in Chapter 3 on autoencoders, in which large models were defined as a set of linked sub-models. While we will not be defining models for each component, we will need to define functions that automatically create new components of networks for us.

- *Automation*: We will define methods of building neural networks that can build more segments of the architecture than we explicitly define in the code, allowing us to scale the network more easily and to develop complex topological patterns.

- *Parametrization*: Networks that explicitly define key architectural parameters like the width and depth are not robust nor scalable. In order to compartmentalize and automate neural network design, we will need to define parameters relative to other parameters rather than as absolute, static values.

You'll find that the content discussed in this chapter may feel more abstracted than that of previous chapters. This is part of the process of developing more complex architectures and a deeper feel for the dynamics of neural networks: rather than seeing an architecture as merely a collection of layers arranged in a certain format that somewhat arbitrarily and magically form a successful prediction function, you begin to identify and use motifs, residuals, parallel branches, cardinality, cells, scaling factors, and other architectural patterns and tools. This eye for design will be invaluable in the development of more professional, successful, and universal network architecture designs.

Nonlinear and Parallel Representation

The realization that purely linear neural network architecture topologies perform well but not well enough drives nonlinearity in the topology of – more or less – every successful modern neural network design. Linearity in the topology of a neural network becomes a burden when we want to scale it to obtain greater modeling power.

A good conceptual model for understanding the success and principles of nonlinear architecture topology design is to think of each layer as its own "thinker" engaged in a larger dialogue – one part of a large, connected network. Passing an input into the network is like presenting this set of thinkers with an unanswered question to consider. Each thinker sheds their unique light on the input, making some sort of transformation to it – reframing the question or adding some progress in answering it, perhaps – before passing the fruits of their consideration onto the next thinker. You want to design an arrangement of these thinkers – that is, where certain thinkers get and pass on information – such that the output of the dialogue (the answer to your input question) takes advantage of each of these thinkers' perspectives as much as possible.

Let's consider how a hypothetical arranged network of thinkers would approach the age-old question, "What is the meaning of life?" (Figure 6-2). The first thinker reframes the question of meaning as a matter of value and focuses on those that possess life – living beings – as the subjects of the question, in turn asking: "What do living beings value most in their life?" The next thinker interprets this question in an anthropocentric sense as relating to how *humans* value life. The last thinker answers that humans value happiness the most, and the output of this network is the common answer of "happiness."

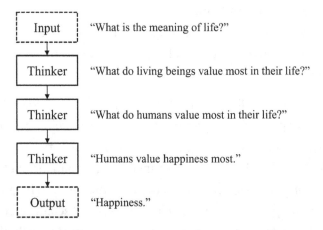

Figure 6-2. *Hypothetical dialogue of a linearly arranged chain of thinkers – an answer to the age-old question, "What is the meaning of life?"*

Because of the linear topology of this arrangement of thinkers, each thinker can only think through the information given to them by the previous thinker. All thinkers after the first thinker don't have direct access to the original input question; the third thinker, for instance, is forced to answer the second thinker's interpretation on the first thinker's interpretation on the original question. While increasing the depth of the network can be incredibly powerful, it can also lead to this problem we see before us that later "thinkers" or layers are progressively detached from the original context of the input.

By adding nonlinearity to our arrangement of thinkers, the network is able to generate more complexity of representation by more directly connecting "thinkers" with the developments of multiple thinkers at various locations down the "chain" of dialogue (Figure 6-3). In this example nonlinear arrangement of thinkers, we add one additional connection from the first thinker to the third thinker such that the third thinker takes in the ideas of both the first and the second thinker. Rather than responding to only the interpretation of the second thinker, it is able to consider the developments and ideas of multiple thinkers.

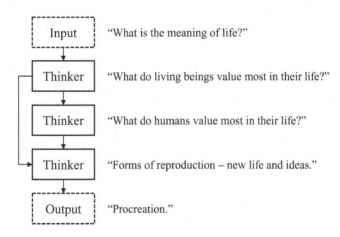

Figure 6-3. *Adding a nonlinearity in the arrangement of thinkers significantly changes the output of the dialogue*

In this example, the third thinker generalizes the developments of the first and second thinkers – "What do living beings value most in their life?" and "What do humans value most in their life?" – between a purely biological perspective on the value of life and another more embedded in the humanities. The third thinker responds: "Forms of reproduction – new life and ideas," referring to the role of reproduction both in biological evolution and in the reproduction of ideas across individuals, generations, and societies of human civilization. To summarize the third thinker's ideas, the output of the network is thus "procreation."

With the addition of one connection from the first thinker to the third thinker, the output of the network has changed from "happiness" – a simpler, more instinctive answer – to the concept of "procreation," a far deeper and more profound response. The key difference here is the element of merging different perspectives and ideas in consideration.

Most people would agree that more, not less, dialogue is conducive to the emergence of great ideas and insights. Similarly, think of each connection as a "one-way conversation." When we add more connections between thinkers, we add more conversations into our network of thinkers. With sufficient number and nonlinearity connections, our network will burst into a flurry of activity, discourse, and dialogue, performing better than a linearly arranged network ever could.

Residual Connections

Residual connections are the first step toward nonlinearity – these are simple connections placed between nonadjacent layers. They're often presented as "skipping" over a layer or multiple layers, which is why they are also often referred to as "skip connections" (Figure 6-4).

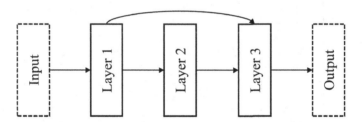

Figure 6-4. *Residual connection*

Note that because the Keras Functional API cannot define multiple layers as the input to another layer, in implementation the connections are first merged through a method like adding or concatenation. The merged components are then passed into the next layer (Figure 6-5). This is the implicit assumption of all residual connection diagrams.

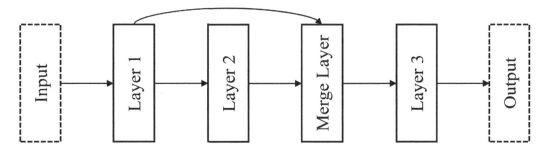

Figure 6-5. *Technically correct residual connection with merge layer*

Although residual connections are incredibly versatile tools, there are generally two methods of residual connection usage, based on the ResNet and DenseNet architectures that pioneered their respective usages of residual connections. "ResNet-style" residual connections employ a series of short residual connections that are repeated routinely throughout the network (Figure 6-6).

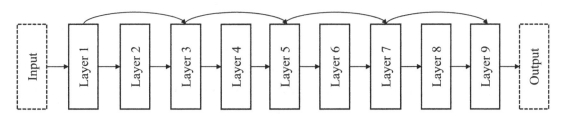

Figure 6-6. *ResNet-style usage of residual connections*

"DenseNet-style" residual connections, on the other hand, place residual connections between valid layers (i.e., layers designated as open to having residual connections attached, also known as "anchor points" or "anchor layers") (Figure 6-7). Because – as one might imagine – this sort of usage of residual connections leads to a high number of residual connections, seldom do DenseNet-style architectures treat all layers as residual connections. In this style, both long and short residual connections are used to provide information pathways across sections of the architecture. Because each anchor point is connected to each anchor point before it, it is informed by various stages of the network's processing and feature extraction.

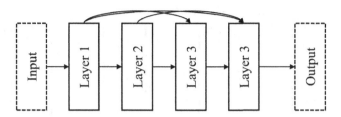

Figure 6-7. *DenseNet-style usage of residual connections*

This representation of residual connections (Figure 6-4) is probably the most convenient interpretation of what a residual connection is for the purposes of implementation. It visualizes residual connections as nonlinearities added upon a "main sequence" of layers – a linear "backbone" upon which residual connections are added. In the Keras Functional API, it's easiest to understand a residual connection as "skipping" over a "main sequence" of layers. In general, implementing nonlinearities is best performed with this concept of a linear backbone in mind, because code is written in a linear format that can be difficult to translate nonlinearities to.

However, there are other architectural interpretations of what a residual connection is that may be conceptually more aligned with the general class of nonlinearities in neural network architectures. Rather than relying upon a linear backbone, you can interpret a residual connection as splitting the layer before it into two branches, which each process the previous layer in their own unique ways. One branch (Layer 1 to Layer 2 to Layer 3 in Figure 6-8) processes the output of the previous layer with a specialized function, whereas the other branch (Layer 1 to Identity to Layer 3) processes the output of the layer with the identity function – that is, it simply allows the output of the previous layer to pass through, the "simplest" form of processing (Figure 6-8).

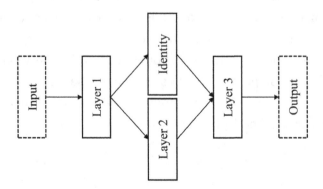

Figure 6-8. *Alternative interpretation of residual connection as a branching operation*

This method of conceptually understanding residual connections more easily allows you to categorize them as a sub-class of general nonlinear architectures, which can be understood as a series of branching structures (Figure 6-9). We'll see how this interpretation helps later in our exploration of parallel branches and cardinality.

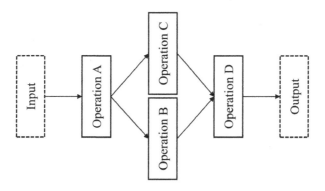

Figure 6-9. *Generalized nonlinearity form*

Residual connections are often presented as a technical justification for the *vanishing gradient problem* (Figure 6-10), which shares many characteristics to the previously discussed problem of a linear arrangement of thinkers: in order to access some layer, we need to travel through several other layers first, diluting the information signal. In the vanishing gradient problem, the backpropagation signal within very deep neural networks used to update the weights gets progressively weaker such that the front layers are barely utilized at all.

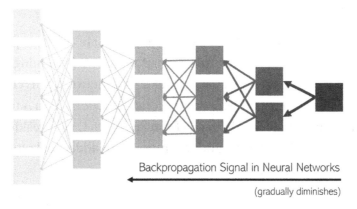

Backpropagation Signal in Neural Networks

(gradually diminishes)

Figure 6-10. *Vanishing gradient problem*

With residual connections, however, the backpropagation signal travels through fewer average layers to reach some particular layer's weights for updating. This enables a stronger backpropagation signal that is better able to make use of the entire model architecture.

There are other interpretations of residual connections too. Like the random forest algorithm is constructed of many smaller decision tree models trained on part of the dataset, a neural network with a sufficient number of residual connections can be thought to be an "ensemble" of smaller sequential models built with a fewer number of layers (Figure 6-11).

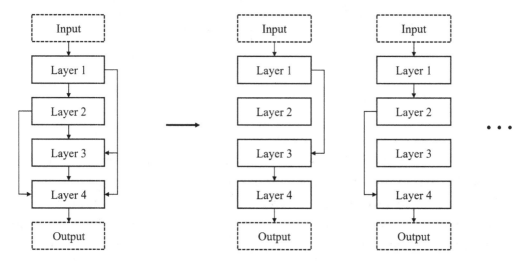

Figure 6-11. *Deconstructing a DenseNet-style network into a series of linear topologies*

Residual connections can also be thought of as a "failsafe" for poor performing layers. If we add a residual connection from layer A to layer C (assuming layer A is connected to layer B, which is connected to layer C), the network can "choose" to disregard layer B by learning near-zero weights for connections from A to B and from B to C while information is channeled directly from layer A to C via a residual connection. In practice, however, residual connections act more as an additional representation of the data for consideration than as failsafe mechanisms.

Implementing individual residual connections is quite simple with our knowledge of the Functional API. We'll use the first presented interpretation of the residual connection architecture, in which a residual connection acts as a "skipping mechanism" between nonadjacent layers in a linear architecture backbone. For simplicity, let's define this linear architecture backbone as a series of Dense layers (Listing 6-1, Figure 6-12).

Listing 6-1. Creating a linear architecture using the Functional API to serve as the linear backbone for residual connections

```
inp = L.Input((128,))
layer1 = L.Dense(64, activation='relu')(inp)
layer2 = L.Dense(64, activation='relu')(layer1)
layer3 = L.Dense(64, activation='relu')(layer2)
output = L.Dense(1, activation='sigmoid')(layer3)
model = keras.models.Model(inputs=inp, outputs=output)
```

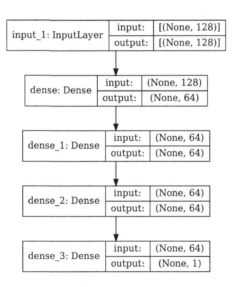

Figure 6-12. *Architecture of a sample linear backbone model*

Let's say that we want to add a residual connection from layer1 to layer3. In order to do this, we need to merge layer1 with whatever current layer is the input to layer3 (this is layer2). The result of the merging is then passed as the input to layer3 (Listing 6-2, Figure 6-13).

Listing 6-2. Building a residual connection by adding a merging layer

```
inp = L.Input((128,))
layer1 = L.Dense(64, activation='relu')(inp)
layer2 = L.Dense(64, activation='relu')(layer1)
concat = L.Concatenate()([layer1, layer2])
layer3 = L.Dense(64, activation='relu')(concat)
output = L.Dense(1, activation='sigmoid')(layer3)
model = keras.models.Model(inputs=inp, outputs=output)
```

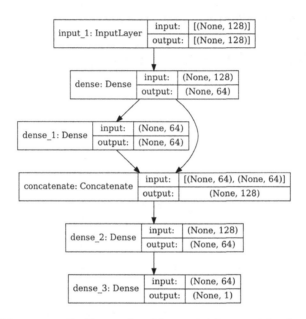

Figure 6-13. *Architecture of a linear backbone with a residual connection skipping over a processing layer, dense_1*

However, this method of manually creating residual connections by explicitly defining merging and input flows is inefficient and not scalable (i.e., it becomes too unorganized and difficult if you wanted to define dozens or hundreds of residual connections). Let's create a function, make_rc() (Listing 6-3), that takes in a connection split layer (this is the layer that "splits" – there are two connections stemming from it) and a connection joining head (this is the layer before the layer in which the residual connection and the "linear main sequence" join together) and outputs a merged version of those two layers that can be used as an input to the next layer (Figure 6-14).

We'll see how automating the creation of residual connections will be incredibly helpful soon when we attempt to construct more elaborate ResNet- and DensNet-style residual connections.

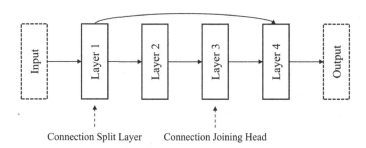

Figure 6-14. *Terminology for certain layers' relationship to one another in the automation of building residual connections*

We can also add another parameter, merge_method, which allows the function user to specify which method to use to merge the connection split layer and the connection joining head. The parameter takes in a string, which is mapped to the corresponding merging layer via a dictionary.

Listing 6-3. Automating the creation of a residual connection by defining a function to create a residual connection

```
def make_rc(split_layer, joining_head,
            merge_method='concat'):
    method_dic = {'concat':L.Concatenate(),
                  'add':L.Add(),
                  'avg':L.Average(),
                  'max':L.Maximum()}
    merge = method_dic[merge_method]
    conn_output = merge([split_layer, joining_head])
    return conn_output
```

We can thus build a residual connection quite easily simply by passing in the function make_rc with the appropriate parameters as an input to the layer we want to receive the merged result (Listing 6-4).

Listing 6-4. Using the residual-connection-making function in a model architecture. (Activation functions may not be present in many listings due to space.)

```
inp = L.Input((128,))
layer1 = L.Dense(64)(inp)
layer2 = L.Dense(64)(layer1)
layer3 = L.Dense(64)(make_rc(layer1, layer2))
output = L.Dense(1)(layer3)
model = keras.models.Model(inputs=inp, outputs=output)
```

We can automate the usage of this function to create a ResNet-style architecture in which blocks of layers with short residual connections are repeated several times (Figure 6-15). In order to automate the construction of the architecture, we will use the placeholder variables x, x1, and x2. We will build x1 from x and x2 from x1 in sequences and merge x with x2 to build the residual connection (Listing 6-5).

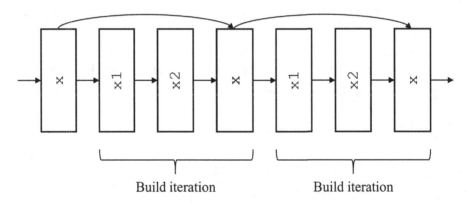

Figure 6-15. *Automating the construction of residual connections by constructing in entire build iterations/blocks*

Tip   It can be conceptually difficult to automate the building of these nonlinear topologies. Drawing a diagram and labeling with which template variables allows you to implement complex residual connection patterns more easily.

Listing 6-5. Building ResNet-style residual connections

```
# number of residual connections
num_rcs = 3

# define input + first dense layer
inp = L.Input((128,))
x = L.Dense(64)(inp)

# create residual connections
for i in range(num_rcs):

    # build two layers to skip over
    x1 = L.Dense(64)(x)
    x2 = L.Dense(64)(x1)

    # define x as merging of x and x2
    x = L.Dense(64)(make_rc(x,x2))
```

Since at the end of the building iterations x is the last connected layer, we connect x to the output layer and aggregate the architecture into a model (Listing 6-6, Figure 6-16).

Listing 6-6. Building output and aggregating ResNet-style architecture into a model

```
# build output
output = L.Dense(1, activation='sigmoid')(x)

# aggregate into model
model = keras.models.Model(inputs=inp, outputs=output)
```

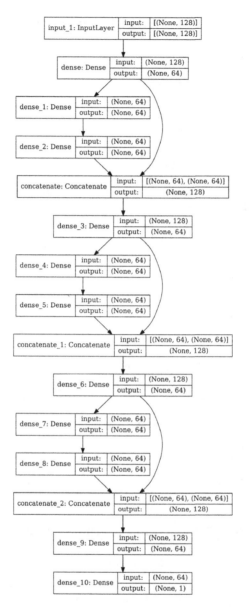

Figure 6-16. *Automated ResNet-style usage of residual connections*

A DenseNet residual connection pattern (Figure 6-17), in which every "anchor point" is connected to every other "anchor point," requires some more planning. (Note that for the sake of ease, we will build residual connections between every Dense layer rather than building additional un-connected layers. We'll discuss residual connections for cell-based architectures in the next section.) We'll keep a list of previous layers x.

Every building step, we add a new layer x[i] that is connected to the merging of every previous layer. Note that the layer x[i-1] is the direct connection and the connection between x[i-2], x[i-3], ... and x[i] is a residual connection.

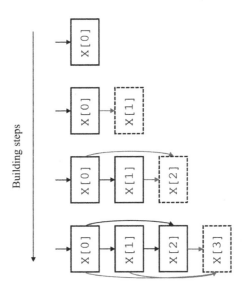

Figure 6-17. *Conceptual diagram of automating the construction of DenseNet-style usage of residual connections*

The first step is to adjust the make_rc function to take in a *list* of layers to join rather than just two. This is because the DenseNet architecture is built such that many residual connections connect to the same layer. In addition to taking in the list of layers, we're going to specify that if the number of layers to join is 1 (i.e., connecting x[0] to x[1] with no residual connections), we simply return the one element in the join_layers list – that is, if there is only one item in a list of layers to merge, we "merge" the layer with an "empty" layer (Listing 6-7).

Listing 6-7. Adjusting the residual-connection-making function for DenseNet-style residual connections

```
def make_rc(join_layers=[],
            merge_method='concat'):
    if len(join_layers) == 1:
        return join_layers[0]
```

```
    method_dic = {'concat':L.Concatenate(),
                  'add':L.Add(),
                  'avg':L.Average(),
                  'max':L.Maximum()}
    merge = method_dic[merge_method]
    conn_output = merge(join_layers)
    return conn_output
```

We can begin by defining an initial layer x and a list of created layers layers. After creating the initial layer, we loop through the remaining layers to be added by redefining the template variable x as a Dense layer that takes in a merged version of all the other currently created layers (Listing 6-8, Figure 6-18). Afterward, we append x to layers such that the next created layer will take in this just created layer.

Listing 6-8. Using the augmented residual-connection-making function to create DenseNet-style residual connections

```
# define number of Dense layers
num_layers = 5

# create input layer
inp = L.Input((128,))
x = L.Dense(64, activation='relu')(inp)

# set layers list
layers = [x]

# loop through remaining layers
for i in range(num_layers-1):

    # define new layer
    x = L.Dense(64)(make_rc(layers))

    # add layer to list of layers
    layers.append(x)

# add output
output = L.Dense(1, activation='sigmoid')(x)
```

```
# build model
model = keras.models.Model(inputs=inp, outputs=output)
```

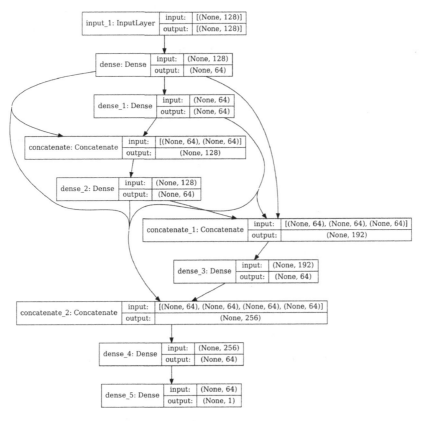

Figure 6-18. *Keras visualization of DenseNet-style residual connections*

This is a great example of how using predefined functions, storage, and built-in scalability in combination with one another allows for a quick and efficient building of complex topologies.

As you can see, when we begin to build more complex network designs, the Keras plotting utility starts to struggle to produce an architecture visualization that is visually consistent with our conceptual understanding of the topology. Regardless, visualization still serves as a sanity check tool to ensure that your automation of complex residual connection relationships is functioning.

Branching and Cardinality

The concept of cardinality is the core framework of nonlinearity and a generalization of residual connections. While the *width* of a network section refers to the number of neurons in the corresponding layer(s), the cardinality of a network architecture refers to the number of "branches" (also known as parallel towers) in a nonlinear space at a certain location in a neural network architecture.

Cardinality is most clear in *parallel branches* – an architectural design in which a layer is "split" into multiple layers, each of which are processed linearly and eventually merged back together. The cardinality of a segment of a network employing parallel branches is simply the number of branches (Figure 6-19).

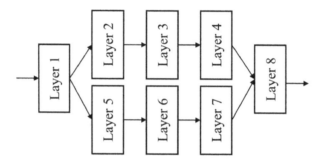

Figure 6-19. *A sample component of a neural network architecture with a cardinality of two. The numbering of the layers is arbitrary*

As mentioned prior in the section on residual connections, one can generalize residual connections as a simple branching mechanism with a cardinality of two, in which one branch is the series of layers the residual connection "skips over" and the other branch is the identity function.

Note that, depending on the specific topology, the cardinality of a section of a network is more or less ambiguous. For instance, some topologies may build branches within sub-branches and join together certain sub-branches in complex ways (e.g., Figure 6-20). Here, the specific cardinality of the network is not relevant; what *is* relevant is that information is being modeled in a nonlinear fashion that encourages multiplicity of representation (the general concept of cardinality) and thus greater complexity and modeling power.

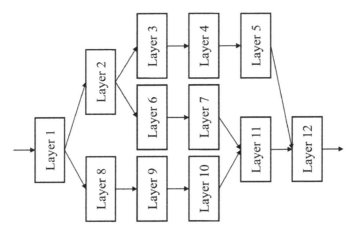

Figure 6-20. *An architectural component of a neural network that demonstrates more extreme nonlinearity (i.e., branching and merging)*

It should be noted, though, that nonlinearities should not be built arbitrarily. It can be easy to get away with building complex nonlinear topologies for the sake of nonlinearity, but you should have some idea of what purpose the topology serves to enable. Likewise, a key component of the network design is not just the architecture within which layers reside but which layers and parameters are chosen to fit within the nonlinear architecture. For instance, you may allocate different branches to use different kernel sizes or to perform different operations to encourage multiplicity of representation. See the case study in the next section on cell-based design for an example on good purposeful design of more complex nonlinearities, the Inception cell.

The logic of branch representations and cardinality in architectural designs is very similar to that of residual connections. This sort of representation is the natural architectural development as a generalization of the residual connection – rather than passing the information flowing through the residual connection (i.e., doing the "skipping") through an identity function, it can be processed separate from other components of the network.

In our analogy of a network as a linked set of thinkers engaged in a dialogue whose net output is dependent on their arrangement, branch representations allow not only for thinkers to consider multiple perspectives but for entire "schools of thought" to emerge in conversation with one another (Figure 6-21). Branches process information separately (i.e., in parallel), allowing different modes of feature extraction to become "mature" (fully formed by several layers of processing) before they are merged with other branches for consideration.

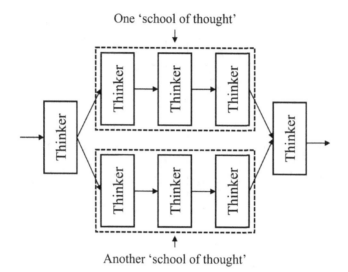

Figure 6-21. *Parallel branches as conceptually organizing "thinkers" into "schools of thought"*

Let's begin by creating a multi-branch nonlinear topology explicitly/manually (Listing 6-9). We will create two branches that span from a layer; each branch, which holds a particular representation of the data, is processed independently and merged later (Figure 6-22).

Listing 6-9. Creating parallel branches manually

```
inp = L.Input((128,))
layer1 = L.Dense(64)(inp)
layer2 = L.Dense(64)(layer1)

branch1a = L.Dense(64)(layer2)
branch1b = L.Dense(64)(branch1a)
branch1c = L.Dense(64)(branch1b)

branch2a = L.Dense(64)(layer2)
branch2b = L.Dense(64)(branch2a)
branch2c = L.Dense(64)(branch2b)

concat = L.Concatenate()([branch1c, branch2c])
layer3 = L.Dense(64, activation='relu')(concat)
```

```
output = L.Dense(1, activation='sigmoid')(layer3)

model = keras.models.Model(inputs=inp, outputs=output)
```

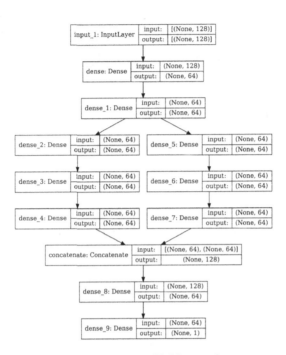

Figure 6-22. *Architecture of building parallel branches*

To automate the building of multiple branches, we consider two dimensions: the number of layers in each branch and the number of parallel branches (Listing 6-10). We can accomplish this in an organized fashion with two functions – build_branch(), which takes in an input layer to start a branch from and outputs the last layer in the branch, and build_branches(), which takes in a layer to split into several branches (using the build_branch() function). We can define the building of a branch simply as a series of linearly connected Dense layers, although branches can also be built as a nonlinear topology too.

Listing 6-10. Automating the building of an individual branch

```
def build_branch(inp_layer, num_layers=5):
    x = L.Dense(64)(inp_layer)
    for i in range(num_layers-1):
        x = L.Dense(64)(x)
    return x
```

In order to split a layer into a series of parallel branches, we use build_branch from that starting layer a certain number of times (this number being the cardinality, which is passed as an argument into the build_branches function). The build_branch function outputs the last layer of the branch, which we append to a list of branch last layers, outputs. After all the branches are built, we merge the outputs together via adding (rather than concatenating in this case, which would yield in a very large concatenated vector) and return the output of merging (Listing 6-11).

Listing 6-11. Automating the building of a series of parallel branches

```
def build_branches(splitting_head, cardinality=4):
    outputs = []

    for i in range(cardinality):
        branch_output = build_branch(splitting_head)
        outputs.append(branch_output)

    merge = L.Add()(outputs)
    return merge
```

Building an entire series of parallel branches within a neural network architecture is now incredibly simple (Listing 6-12, Figure 6-23).

Listing 6-12. Building parallel branches into a complete model architecture

```
inp = L.Input((128,))
layer1 = L.Dense(64)(inp)
layer2 = L.Dense(64)(layer1)
layer3 = L.Dense(64)(build_branches(layer2))
output = L.Dense(1)(layer3)
model = keras.models.Model(inputs=inp, outputs=output)
```

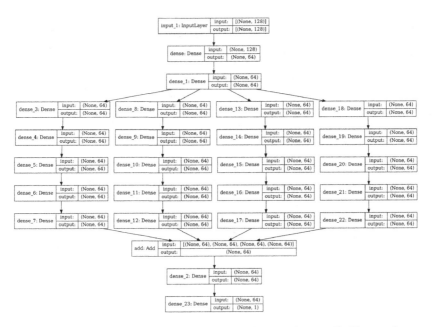

Figure 6-23. *Automation of building arbitrarily sized parallel branches*

This sort of method is used by the ResNeXt architecture, which employs parallel branches as a generalization and "next step" from residual connections.

Case Study: U-Net

The goal of *semantic segmentation* is to segment, or separate, various items within an image into various classes. Semantic segmentation can be used to identify items in a picture, like cars, people, and buildings in a picture of the city. The difference between semantic segmentation and a task like image recognition is that semantic segmentation is an *image-to-image* task, whereas image recognition is an *image-to-vector* task. Image recognition tells you if an object is present in the image or not; semantic segmentation tells you where the object is located in the image by marking each pixel as part of the corresponding class or not. The output of semantic segmentation is called the *segmentation map*.

Semantic segmentation has many applications in biology, which can be used to automate the identification of cells, organs, neuron links, and other biological entities (Figure 6-24). As such, much of research into semantic segmentation architectures has been developed with these biological applications in mind.

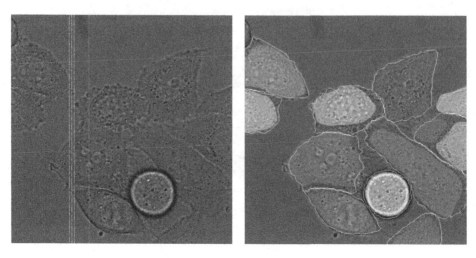

Figure 6-24. *Left: input image to a segmentation model. Right: example segmentation of cells in the image. Taken from the U-Net paper by Ronneberger et al.*

Olaf Ronneberger, Philipp Fischer, and Thomas Brox proposed the *U-Net architecture* in 2015,[1] which has since become a pillar of semantic segmentation development (Figure 6-25). The linear backbone component of the U-Net architecture acts like an autoencoder – it successively reduces the dimension of the image until it reaches some minimum representation size, upon which the dimension of the image is successively increased via upsampling and up-convolutions. The U-Net architecture employs very large residual connections that connect large parts of the network together, connecting the first block of layers to the last block, the second block to the second-to-last block, etc. When the residual connections are arranged to be parallel to one another in an architectural diagram, the linear backbone is forced into a "U" shape, hence its name.

---

[1] Olaf Ronneberger, Philipp Fischer, and Thomas Brox, "Convolutional Networks for Biomedical Image Segmentation," 2015. Paper link: https://arxiv.org/pdf/1505.04597.pdf.

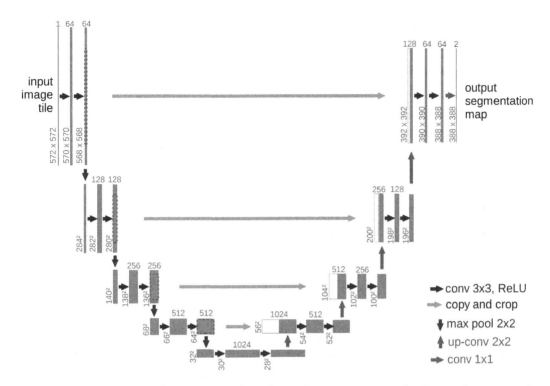

Figure 6-25. *U-Net architecture. Taken from the U-Net paper by Ronneberger et al.*

The left side of the architecture is termed the *contracting path*; it develops successively smaller representations of the data. On the other hand, the right side is termed the *expansive path*, which successively increases the representation of the data. Residual connections allow the network to incorporate previously processed representations into the expansion of the representation size. This architecture allows both for *localization* (focus on a local region of the data input) and the use of broader *context* (information from farther, nonlocal regions of the image that still provide useful data).

The resulting U-Net architecture performed significantly better than other architectures at its time on various biological segmentation challenges (Table 6-1).

Table 6-1. *Performance of U-Net against other segmentation models at the time on two key datasets in the ISBI cell tracking challenge 2015. Results are measured in IOU (Intersection Over Union), a segmentation metric measuring how much of the predicted and true segmented regions overlap. A higher IOU is better. U-Net beats other methods at the time by a large margin in IOU*

Model	PhC-U373 Dataset	DIC-HeLa Dataset
IMCB-SG (2014)	0.2669	0.2935
KTH-SE (2014)	0.7953	0.4607
HOUS-US (2014)	0.5323	–
Second Best (2015)	0.83	0.46
U-Net (2015)	0.9203	0.7756

While the U-Net architecture is technically adaptable to all image sizes because it is built using only convolution-type layers that do not require a fixed input, the original implementation by Ronneberger et al. requires significant planning with respect to the shape of information as it passes through the network. Because the proposed U-Net architecture does not use padding, the spatial resolution is successively reduced and increased divisively (by max pooling) and additively (by convolutions). This means that residual connections must include a cropping function in order to properly merge earlier representations with later representations that have a smaller spatial size. Moreover, the input shape must be carefully planned to have a certain input size, which will not match the output.

Thus, in our implementation of the U-Net architecture, we will slightly adapt the convolutional layers to use padding such that merging and keeping track of data shapes is simpler. While implementing U-Net is relatively simple, it's important to keep track of variables. We will name our layers with appropriate variable names such that they can be used in the make_rc() function earlier discussed to be part of a residual connection.

Building the contracting path (Listing 6-13) is quite simple; in this case, we implement one convolution before every pooling reduction, although you can add more. We use three pooling reductions; conv4 is the "bottleneck" of the autoencoder, carrying the data with the smallest spatial representation size. Note that we increase the number of filters in each convolutional layer to accommodate for reductions in resolution to avoid actually building a representative bottleneck, which would be counteractive to the goals of this model.

Listing 6-13. Building the contracting path of the U-Net architecture

```
inp = L.Input((256,256,3))

# contracting path
conv1 = L.Conv2D(16, (3,3), padding='same')(inp)
pool1 = L.MaxPooling2D((2,2))(conv1)

conv2 = L.Conv2D(32, (3,3), padding='same')(pool1)
pool2 = L.MaxPooling2D((2,2))(conv2)

conv3 = L.Conv2D(64, (3,3), padding='same')(pool2)
pool3 = L.MaxPooling2D((2,2))(conv3)

conv4 = L.Conv2D(128, (3,3), padding='same')(pool3)
```

Building the expanding path (Listing 6-14) requires some more caution. We begin by upsampling the last layer, conv4, such that it has the same spatial dimension as the conv3 layer. We apply another convolution after upsampling to process the result, as well as to ensure that upsamp4 has the same depth (i.e., number of channels) as upsamp3. They can then be merged together via adding to preserve the depth.

Listing 6-14. Building one component of the expanding path in the U-Net architecture

```
# expanding path
upsamp4 = L.UpSampling2D((2,2))(conv4)
upsamp4 = L.Conv2D(64, (3,3), padding='same')(upsamp4)
merge3 = make_rc([conv3, upsamp4], merge_method='add')
```

We can likewise build the remainder of the expanding path (Listing 6-15).

Listing 6-15. Building the remaining components of the expanding path in the U-Net architecture

```
upsamp3 = L.UpSampling2D((2,2))(merge3)
upsamp3 = L.Conv2D(32, (3,3), padding='same')(upsamp3)
merge2 = make_rc([conv2, upsamp3])
```

```
upsamp2 = L.UpSampling2D((2,2))(merge2)
upsamp2 = L.Conv2D(16, (3,3), padding='same')(upsamp2)
merge1 = make_rc([conv1, upsamp2])
```

To ensure that the input data and the output of the U-Net architecture have the same number of channels, we add a convolution layer with (1,1) that doesn't change the spatial dimension but collapses the number of channels to the standard three used in almost all image data (Listing 6-16).

Listing 6-16. Adding an output layer to collapse channels and aggregating layers into a model

```
out = L.Conv2D(3, (1,1))(merge1)
model = keras.models.Model(inputs=inp, outputs=out)
```

As mentioned earlier, you can change the input shape and the code will produce a valid result, as long as the spatial dimensions of the input are divisible by 2. You can plot the model with plot_model to reveal the architectural namesake of the model (Figure 6-26).

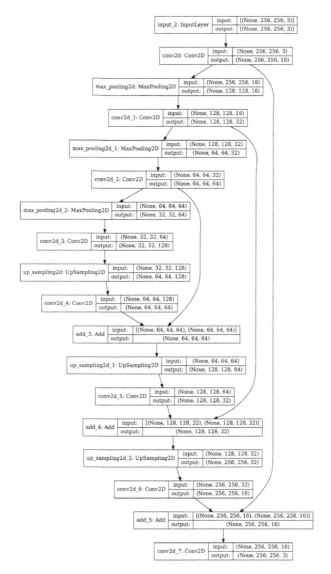

Figure 6-26. *U-Net-style architecture implementation in Keras. Can you see the "U"?*

Block/Cell Design

Recall the network of thinkers used to introduce the concept of nonlinearity. With the addition of a nonlinearity, thinkers were able to consider perspectives from multiple stages of development and processing and therefore expand their worldview to produce more insightful outputs.

357

Profound intellectual work is best achieved via conversation between *multiple thinkers* rather than the processing capability of one individual thinker (Figure 6-27). Thus, it makes sense to consider the key intellectual unit of thinking as a *structured arrangement of thinkers* rather than just one individual thinker. These "cells" are the new unit objects, which can be stacked together like "super-layers" in a similar way that we stacked individual thinkers before – we can stack these cells of thinkers linearly, in branches, add residual connections, etc. By replacing the base unit of information processing with a more powerful unit, we systematically increase the modeling power of the entire system.

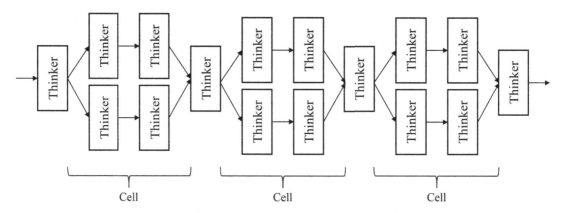

Figure 6-27. *Arrangement of thinkers into cells*

Likewise, by arranging neural network layers into cells, we form a new unit with which we can manipulate architecturally – the cell, which contains more processing capability than a single layer. A cell can be thought of as a "mini-network."

In a block/cell-based design, a neural network consists of several repeating "cells," each of which contains a preset arrangement of layers (Figure 6-28). Cells have shown to be a simple and effective way to increase the depth of a network. These cells form "motifs," which are recurring themes or patterns in the arrangement of certain layers. Like many observed phenomena in deep learning, cell-based design has parallels in neuroscience – neural circuits have been observed to assemble into repeated motif patterns. Cell-based designs allow for established and standardized extraction of features; the top performing neural network architectures include a high number of well-designed cell-based neural network architectures.

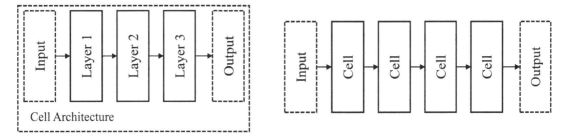

Figure 6-28. *Usage of cell-based design*

From a design standpoint, cells reduce the search space a neural network designer is confronted with. (You may recall that this is the justification given for a NASNet-style search space for cell architectures discussed in Chapter 5.) The designer is able to exercise fine-tuned control over the architecture of the cell, which is stacked repeatedly, therefore amplifying the effect of changes to the architecture. Contrast this with making a change to one layer in a non-cell-based network – it is unlikely that making one change would be significant enough to yield meaningfully different results (provided there exist other similar layers).

There are two key factors to consider in cell-based design: the architecture of the cell itself and the stacking method. This section will discuss methods of stacking across sequential and nonlinear cell design.

Sequential Cell Design

Sequential cell designs are – as implied by the name – cell architectures that follow a sequential, or linear, topology. While sequential cell designs have not generally performed as well as nonlinear cell designs, they can be useful as a beginning location to illustrate key concepts that will be used in nonlinear cell designs.

We'll begin with building a *static dense cell* – a cell whose primary processing layer is the fully connected layer and that does not change its layers to adapt to the input shape (Listing 6-17). Using the same logic as previously established, a function is defined to take in an input layer upon which the cell will be built. The last layer of the cell is then returned to be the input to the next cell (or to the output layer).

Listing 6-17. Building a static dense cell

```
def build_static_dense_cell(inp_layer):
    dense_1 = L.Dense(64)(inp_layer)
    dense_2 = L.Dense(64)(dense_1)
    batchnorm = L.BatchNormalization()(dense_2)
    dropout = L.Dropout(0.1)(batchnorm)
    return dropout
```

These cells can then be repeatedly stacked by iteratively passing the output of the previously constructed cell as the input to the next cell (Listing 6-18). This combination of cell design and stacking – linear cell design stacked in a linear fashion – is perhaps the simplest cell-based architectural design (Figure 6-29).

Listing 6-18. Stacking static dense cells together

```
num_cells = 3

inp = L.Input((128,))
x = build_static_dense_cell(inp)
for i in range(num_cells-1):
    x = build_static_dense_cell(x)
output = L.Dense(1, activation='sigmoid')(x)

model = keras.models.Model(inputs=inp, outputs=output)
```

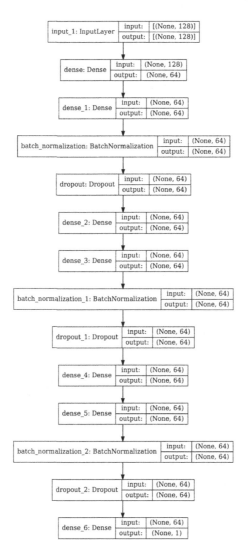

Figure 6-29. *Visualization of cell-based design architecture*

Note that while we are building a cell-based architecture, we're not using *compartmentalized design* (introduced in Chapter 3 on autoencoders). That is, while conceptually we understand that the model is built in terms of repeated block segments, we do not build that into the actual implementation of the model by defining each cell as a separate model. In this case, there is no need for compartmentalized design since we do not need to access the output of any one particular cell. Moreover, defining builder functions should serve as a sufficient code-level conceptual organization of how the network is constructed. The primary burden of using compartmentalized design with cell-based structures is the need to keep track of the input shape of the previous cell's output in an automated fashion, which is required when defining separate models. However, implementing compartmentalized design will make Keras' architecture visualizations more consistent with our conceptual understanding of cell-based models by visualizing cells rather than the entire sequence of layers.

We can similarly create a static convolutional cell by using the standard sequence of a series of convolutional layers followed by a pooling layer, with batch normalization and dropout for good measure (Listing 6-19).

Listing 6-19. Building a static convolutional cell

```
def build_static_conv_cell(inp_layer):
    conv_1 = L.Conv2D(64,(3,3))(inp_layer)
    conv_2 = L.Conv2D(64,(3,3))(conv_1)
    pool = L.MaxPooling2D((2,2))(conv_2)
    batchnorm = L.BatchNormalization()(pool)
    dropout = L.Dropout(0.1)(batchnorm)
    return dropout
```

We can combine the convolutional and dense cells together (Listing 6-20, Figure 6-30). After the convolutional component is completed, we use Global Average Pooling 2D (the Flatten layer also works) to collapse the image-based information flow into a vector-like shape that is processable by the fully connected component.

Listing 6-20. Stacking convolutional and dense cells together in a linear fashion

```
num_conv_cells = 3
num_dense_cells = 2

inp = L.Input((256,256,3))
conv_cell = build_static_conv_cell(inp)
for i in range(num_conv_cells-1):
    conv_cell = build_static_conv_cell(conv_cell)
collapse = L.GlobalAveragePooling2D()(conv_cell)
dense_cell = build_static_dense_cell(collapse)
for i in range(num_dense_cells-1):
    dense_cell = build_static_dense_cell(dense_cell)
output = L.Dense(1, activation='sigmoid')(dense_cell)

model = keras.models.Model(inputs=inp, outputs=output)
```

362

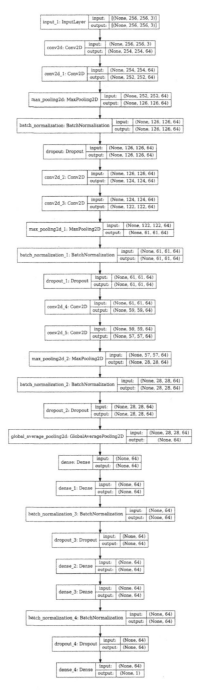

Figure 6-30. *Visualization of architecture stacking convolutional and dense cells*

Note that static dense and static convolutional cells have different effects on the data input shapes they are applied to. Static dense cells always output the same data output shape, since the user specifies the number of nodes the input will be projected to when defining a Dense layer in Keras. On the other hand, static convolutional cells output different data output shapes, depending on which data input shapes they received. (This is discussed in more detail in Chapter 2 on transfer learning.) Because different primary layer types in cell designs can yield different impacts on the data shape, it is general convention not to build static cells.

Instead, cells are generally built by their effect on the output shape, such that they can more easily be stacked together. This is especially helpful in nonlinear stacking patterns, in which cell outputs must match to be merged in a valid fashion. Generally, cells can be categorized as reduction cells or normal cells. Normal cells keep the output shape the same shape as the input shape, whereas reduction cells decrease the output shape from the input shape. Many modern architectures employ multiple designs for normal and reduction cells that are stacked repeatedly throughout the model.

In order to build these shape-based cells (Listing 6-21), we will need an additional parameter to keep track of – the input shape. For a network dealing with tabular data, we are concerned only with the width of the input layer. In a normal cell, the output is the same width as the input; we define a reduction cell to reduce the size of the input by half, although you can adopt different designs, depending on your problem type. Each cell-building function returns the output layer of the cell, in addition to the width of the output layer. This information will be employed in building the next cell.

Listing 6-21. Building normal and reduction cells

```
def build_normal_cell(inp_layer, width):
    dense_1 = L.Dense(width)(inp_layer)
    dense_2 = L.Dense(width)(dense_1)
    return dense_2, width

def build_reduce_cell(inp_layer, width):
    new_width = round(width/2)
    dense_1 = L.Dense(new_width)(inp_layer)
    dense_2 = L.Dense(new_width)(dense_1)
    return dense_2, new_width
```

We can simply sequentially stack these two cells together in an alternating pattern (Listing 6-22, Figure 6-31). We use the holder variables `cell_out` and `w` to keep track of the output of a cell and the corresponding width. Each cell-building function either keeps the shape `w` the same or modifies it to reflect changes in the shape of the output layer of the cell. This information is, as mentioned before, passed into the following cell-building functions.

Listing 6-22. Stacking reduction and normal cells linearly

```
num_repeats = 2

w = 128
inp = L.Input((w,))
cell_out, w = build_normal_cell(inp, w)
for repeat in range(num_repeats):
    cell_out, w = build_reduce_cell(cell_out, w)
    cell_out, w = build_normal_cell(cell_out, w)
output = L.Dense(1, activation='sigmoid')(cell_out)

model = keras.models.Model(inputs=inp, outputs=output)
```

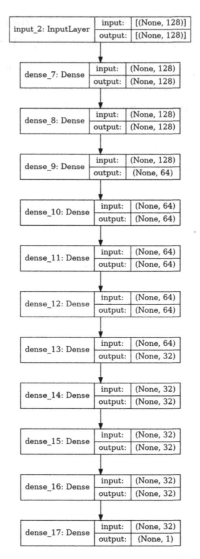

Figure 6-31. *Architecture of alternating stacks of cells*

Building convolutional normal and reduction cells is similar, but we need to keep track of three shape elements rather than one. Like in the context of autoencoders and other shape-sensitive contexts, it's best to use padding='same' to keep the input and output shapes the same.

The normal cell uses two convolutional cells with 'same' padding (Listing 6-23). You can also use a pooling layer with padding='same' if you'd like to use pooling in a normal cell. Note that the depth of the input is passed as the number of filters in the convolutional layers to preserve the input shape.

Listing 6-23. Building a convolutional normal cell

```
def build_normal_cell(inp_layer, shape):

    h,w,d = shape

    conv_1 = L.Conv2D(d,(3,3),padding='same')(inp_layer)
    conv_2 = L.Conv2D(d,(3,3),padding='same')(conv_1)

    return conv_2, shape
```

The reduction cell (Listing 6-24) also uses two convolutional cells with 'same' padding, but a pooling layer is added, which reduces the height and width by half. We return the new shape, which halves the height and width. The ceiling operation is performed on the result, in the case that the input shape height or width is odd and the result of division is not an integer. If you use a different padding mode for max pooling, you'll need to correspondingly adjust how the new shape is calculated.

Listing 6-24. Building a convolutional reduction cell

```
def build_reduce_cell(inp_layer, shape):

    h,w,d = shape

    conv_1 = L.Conv2D(d,(3,3),padding='same')(inp_layer)
    conv_2 = L.Conv2D(d,(3,3),padding='same')(conv_1)
    pool = L.MaxPooling2D((2,2))(conv_2)

    new_shape = (np.ceil(h/2),np.ceil(w/2),d)
    return pool, new_shape
```

These two cells can then be stacked in a linear arrangement, as previously demonstrated. Alternatively, these cells can be stacked in a nonlinear format. To do this, we'll use the methods and ideas developed from the nonlinear and parallel representation section.

To merge the outputs of two cells together, they need to have the same shape. Thanks to shape-based design, we can arrange normal and reduction cells accordingly to form valid merge operations. For instance, we can draw a residual connection from a normal cell "over" another normal cell, merge the two connections, and pass the merged result into a reduction cell as displayed in Figure 6-32.

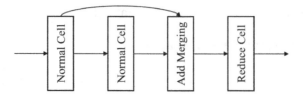

Figure 6-32. *Nonlinear stacking of cells*

However, reduction cells form "borders" that cannot be transgressed by connections (unless you use reshaping mechanisms), because the input shapes on the two sides of the reduction cell do not match (Figure 6-33).

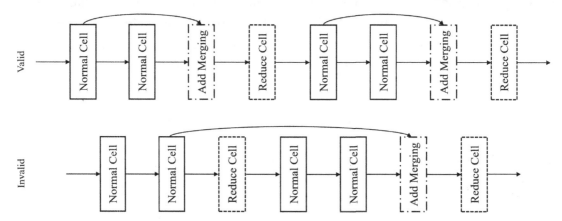

Figure 6-33. *Demonstration of residual connection usage across cells*

We can build nonlinear cell-stacking architectures in a very similar way as we built nonlinearities and parallel representations among layers in a network architecture. We will use the make_rc function defined in the previous section to merge the outputs of norm_1 and norm_2 and pass the merged result as the input of a reduction cell (Listing 6-25, Figure 6-34). The shape is defined after the input with a depth of 64 such that convolutional layers process the data with 64 filters (rather than 3). If desired, you can manipulate reduction cells to also change the image depth. Note that the merging operation we use in this case is adding rather than concatenation. The output shape of adding is the same as the input shape of any one of the inputs, whereas depth-wise concatenation changes the shape. This can be accommodated but requires ensuring the shape is correspondingly updated.

Listing 6-25. Stacking convolutional normal and reduction cells together in nonlinear fashion with residual connections

```
inp = L.Input((128,128,3))
shape = (128,128,64)
norm_1, shape = build_normal_cell(inp, shape)
norm_2, shape = build_normal_cell(norm_1, shape)
merged = make_rc([norm_1,norm_2],'add')
reduce, shape = build_reduce_cell(merged, shape)
```

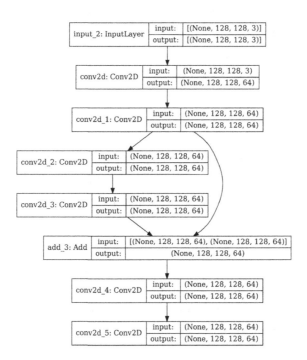

Figure 6-34. *Residual connections across normal cells*

Cells can similarly be stacked in DenseNet style and with parallel branches for more complex representation. If you're using linear cell designs, it's recommended to use nonlinear stacking methods to add nonlinearity into the general network topology.

Nonlinear Cell Design

Nonlinear cell designs have only one input and output, but use nonlinear topologies to develop multiple parallel representations and processing mechanisms, which are combined into an output. Nonlinear cell designs are generally more successful (and hence more popular) because they allow for the construction of more powerful cells.

Building nonlinear cell designs is relatively simple with knowledge of nonlinear representation and sequential cell design. The design of nonlinear cells closely follows the previous discussion on nonlinear and parallel representation. Nonlinear cells form organized, highly efficient feature extraction modules that can be chained together to successively process information in a powerful, easily scalable manner.

Branch-based design is especially powerful in the design of nonlinear cell architectures. It is able to effectively extract and merge different parallel representations of data in one compact, stackable cell.

Let's build a normal cell for image data that keeps the spatial dimensions and depth of the data constant (Listing 6-26, Figure 6-35). Like previous designs, it will take in both the input layer to attach the cell to, as well as the shape of the data. The depth of the image will be extracted from the shape and used as the depth throughout the cell. By using appropriate padding, the spatial dimensions of the data remain unchanged too. Three branches each extract and process features in parallel with different filter sizes; these representations are then merged via concatenation (depth-wise, meaning that they are "stacked together"). This merging produces data of shape $(h, w, d \cdot 3)$, as we are stacking together the outputs of three branches; to ensure that the output shape is identical to the input shape, we add another convolutional layer with filter $(1, 1)$ to collapse the number of channels from $d \cdot 3$ to d.

Listing 6-26. Build a convolutional nonlinear normal cell

```
def build_normal_cell(inp_layer, shape):

    h,w,d = shape

    branch1a = L.Conv2D(d,(5,5),padding='same')(inp_layer)
    branch1b = L.Conv2D(d,(3,3),padding='same')(branch1a)
    branch1c = L.Conv2D(d,(1,1))(branch1b)

    branch2a = L.Conv2D(d,(3,3),padding='same')(inp_layer)
    branch2b = L.Conv2D(d,(3,3),padding='same')(branch2a)
```

```
branch3a = L.Conv2D(d,(3,3),padding='same')(inp_layer)
branch3b = L.Conv2D(d,(1,1))(branch3a)

merge = L.Concatenate()([branch1c, branch2b, branch3b])
out = L.Conv2D(d, (1,1))(merge)

return out, shape
```

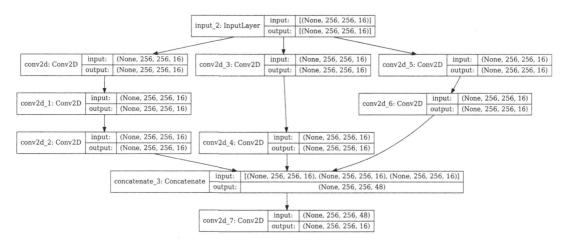

Figure 6-35. *Hypothetical architecture of nonlinear normal cell*

We can similarly build a reduction cell by building multiple branches that reduce the spatial dimension of the input (Listing 6-27, Figure 6-36). In this case, one branch performs a convolution with stride (2, 2) – halving the spatial dimension. Another uses a standard max pooling reduction. Both are followed by a convolution layer with filter (1,1) that separately process the output before they are merged together via depth-wise concatenation. When the two are concatenated, the depth is doubled. This is fine in this case; since we want to compensate for decreases in resolution with corresponding increase in quantity of filters, there is no need to decrease the number of filters after concatenation as performed in the design of the normal cell. Correspondingly, we place a convolutional layer with filter (1,1) to further process the results of concatenation and use that layer as the output. The new shape of the data is correspondingly calculated and passed as a second output to the cell-building function.

Listing 6-27. Building a convolutional nonlinear reduction cell

```
def build_reduction_cell(inp_layer, shape):

    h,w,d = shape

    branch1a = L.Conv2D(d,(3,3), strides=(2,2), padding='same')(inp_layer)
    branch1b = L.Conv2D(d,(1,1))(branch1a)

    branch2a = L.MaxPooling2D((2,2),padding='same')(inp_layer)
    branch2b = L.Conv2D(d,(1,1))(branch2a)

    merge = L.Concatenate()([branch1b, branch2b])
    out = L.Conv2D(d*2, (1,1))(merge)

    new_shape = (np.ceil(h/2), np.ceil(w/2), d*2)
    return out, new_shape
```

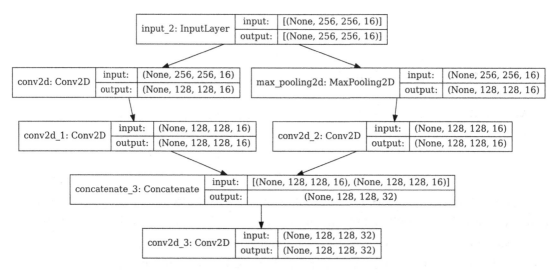

Figure 6-36. *Hypothetical architecture of a nonlinear reduction cell*

These two cells (and any other additional designs for normal or reduction cells) can be stacked linearly. Because nonlinearly designed cells contain sufficient nonlinearity, there is less of a need to be aggressive in nonlinear stacking. Stacking nonlinear cells sequentially is a tried-and-true formula. One concern that could arise with linear stacking arises if you are stacking so many cells together that the depth of the network poses problems for cross-network information flow. Using residual connections across cells can help to address this problem. In the case study for this section, on the famed InceptionV3 model, we will explore a concrete example of successful nonlinear cell-based architectural design.

Case Study: InceptionV3

The famous InceptionV3 architecture, part of the Inception family of models that has become a pillar of image recognition, was proposed by Christian Szegedy, Vincent Vanhoucke, Sergey Ioffe, Jonathon Shlens, and Zbigniew Wojna in the 2015 paper "Rethinking the Inception Architecture for Computer Vision."[2] The InceptionV3 architecture, in many ways, laid out the key principles of convolutional neural network design for the following years to come. The aspect most relevant for this context is its cell-based design.

The InceptionV3 model attempted to improve upon the designs of the previous InceptionV2 and original Inception models. The original Inception model consisted of a series of repeated cells (referred to in the paper as "modules") that followed a multi-branch nonlinear architecture (Figure 6-37). Four branches stem from the input to the module; two branches consist of a 1x1 convolution followed by a larger convolution, one branch is defined as a pooling operation followed by a 1x1 convolution, and another is just a 1x1 convolution. Padding is provided on all operations in these modules such that the size of the filters is kept the same such that the results of the parallel branch representations can be concatenated depth-wise back together.

---

[2] Christian Szegedy, Vincent Vanhoucke, Sergey Ioffe, Jonathon Shlens, and Zbigniew Wojna, "Rethinking the Inception Architecture for Computer Vision," 2015. Paper link: https://arxiv.org/pdf/1512.00567.pdf.

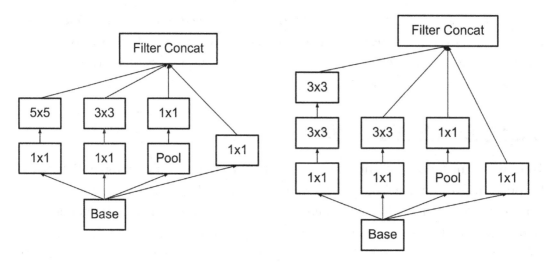

Figure 6-37. *Left: original Inception cell. Right: one of the InceptionV3 cell architectures*

A key architectural change in the InceptionV3 module designs is the *factorization* of large filter sizes like 5x5 into a combination of smaller filter sizes. For instance, the effect of a 5x5 filter can be "factored" into a series of two 3x3 filters; a 5x5 filter applied on a feature map (with no padding) yields the same output shape as two 3x3 filters: (w-4, h-4, d). Similarly, a 7x7 filter can be "factored" into three 3x3 filters. Szegedy et al. note that this factorization does not decrease representative power while promoting faster learning. This module will be termed the *symmetric factorization module*, although in implementation within the context of the InceptionV3 architecture it is referred to as *Module A*.

In fact, even 3x3 and 2x2 filters can be factorized into sequences of convolutions with smaller filter sizes. An *n* by *n* convolution can be represented as a 1 by *n* convolution followed by an *n* by 1 convolution (or vice versa). Convolutions with kernel height and widths that are different lengths are known as *asymmetric convolutions* and can be valuable fine-grained feature detectors (Figure 6-38). In the InceptionV3 module architecture, *n* was chosen to be 7. This module will be termed the *asymmetric factorization module* (also known as *Module B*). Szegedy et al. find that this module performs poorly on early layers but works well on medium-sized feature maps; it is correspondingly placed after symmetric factorization modules in the InceptionV3 cell stack.

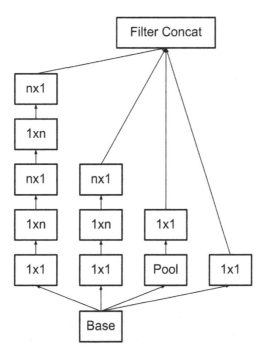

Figure 6-38. *Factorizing n by n filters as operations of smaller filters*

For extremely *coarse* (i.e., small-sized) inputs, a different module with *expanded filter bank outputs* is used. This model architecture encourages the development of high-dimensional representations by using a tree-like topology – the two left branches in the symmetric factorization module are further "split" into "child nodes," which are concatenated along with the outputs of the other branches at the end of the filter (Figure 6-39). This type of module is placed at the end of the InceptionV3 architecture to handle feature maps when they have become small spatially. This module will be termed the *expanded filter bank module* (or *Module C*).

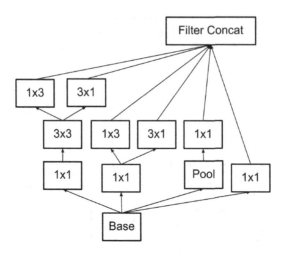

Figure 6-39. *Expanded filter bank cell – blocks within the cell are further expanded by branching into other filter sizes*

Another reduction-style Inception module is designed to efficiently reduce the size of the filters (Figure 6-40). The reduction-style module uses three parallel branches; two use convolutions with a stride of 2 and the other uses a pooling operation. These three branches produce the same output shapes, which can be concatenated depth-wise. Note that Inception modules are designed such that a decrease in size is correspondingly counteracted with an increase in the number of filters.

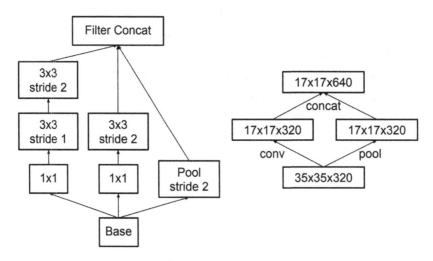

Figure 6-40. *Design of a reduction cell*

The InceptionV3 architecture is formed by stacking these module types in a linear fashion, ordered such that each module is placed in a location where it will receive a feature map input shape that it succeeds in processing. The following sequence of modules is used:

1. A series of convolutional and pooling layers to perform initial feature extraction (these are not part of any module).

2. Three repeats of the symmetric convolution module/Module A.

3. Reduction module.

4. Four repeats of the asymmetric convolution module/Module B.

5. Reduction module.

6. Two repeats of the expanded filter bank module/Module C.

7. Pooling, dense layer, and softmax output.

Another often unnoticed but important feature of the Inception family of architectures is the 1x1 convolution, which is present in every Inception cell design, often as the most frequently occurring element in the architecture. As discussed in previous chapters, using convolutions with kernel size (1,1) is convenient in terms of building architectures like autoencoders when there is a need to collapse the number of channels. In terms of model performance, though, 1x1 convolutions serve a key purpose in the Inception architecture: computing cheap filter reductions before expensive, larger kernels are applied to feature map representations. For instance, suppose at some location in the architecture 256 filters are passed into a 1x1 convolutional layer; the 1x1 convolutional layer can reduce the number of filters to 64 or even 16 by learning the optional combination of values for each pixel from all 256 filters. Because the 1x1 kernel does not incorporate any spatial information (i.e., it doesn't take into account pixels next to one another), it is cheap to compute. Moreover, it isolates the most important features for the following larger (and thus more expensive) convolution operations that incorporate spatial information.

Via well-designed module architectures and purposefully planned arrangement of modules, the InceptionV3 architecture performed very well in that year's ILSVRC (ImageNet competition) and has become a staple in image recognition architectures (Tables 6-2 and 6-3).

Table 6-2. *Performance of InceptionV3 architecture against other models in ImageNet*

Architecture	Top 5 Error	Top 1 Error
GoogLeNet	–	9.15%
VGG	–	7.89%
Inception	22%	5.82%
PReLU	24.27%	7.38%
InceptionV3	18.77%	4.2%

Table 6-3. *Performance of an ensemble of InceptionV3 architectures compared against ensembles of other architecture models*

Architecture	# Models	Top 5 Error	Top 1 Error
VGGNet	2	23.7%	6.8%
GoogLeNet	7	–	6.67%
PReLU	–	–	4.94%
Inception	6	20.1%	4.9%
InceptionV3	4	17.2%	3.58%

The full InceptionV3 architecture is available at `keras.applications.InceptionV3` with available ImageNet weights for transfer learning or just as a powerful architecture (used with random weight initialization) for image recognition and modeling.

Building an InceptionV3 module itself is pretty simple, and because the design of each cell is relatively small, there is no need to automate its construction. We can build four branches in parallel to one another, which are concatenated. Note that we specify `strides=(1,1)` in addition to `padding='same'` in the max pooling layer to keep the input and output layers the same. If we only specify the latter, the strides argument is set to the entered pool size. These cells can then be stacked alongside other cells in a sequential format to form an InceptionV3-style architecture (Listing 6-28, Figure 6-41).

Listing 6-28. Building a simple InceptionV3 Module A architecture

```python
def build_iv3_module_a(inp, shape):

    w, h, d = shape

    branch1a = L.Conv2D(d, (1,1))(inp)
    branch1b = L.Conv2D(d, (3,3), padding='same')(branch1a)
    branch1c = L.Conv2D(d, (3,3), padding='same')(branch1b)

    branch2a = L.Conv2D(d, (1,1))(inp)
    branch2b = L.Conv2D(d, (3,3), padding='same')(branch2a)

    branch3a = L.MaxPooling2D((2,2), strides=(1, 1),
                             padding='same')(inp)
    branch3b = L.Conv2D(d, (1,1), padding='same')(branch3a)

    branch4a = L.Conv2D(d, (1,1))(inp)

    concat = L.Concatenate()([branch1c, branch2b,
                             branch3b, branch4a])
    return concat, shape
```

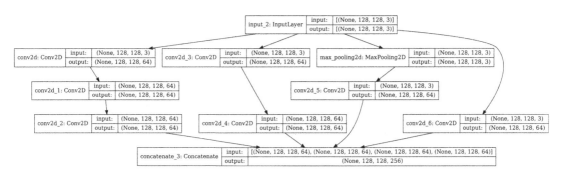

Figure 6-41. *Visualization of a Keras InceptionV3 cell*

Besides getting to work with large neural network architectures directly, another benefit of implementing these sorts of architectures from scratch is customizability. You can insert your own cell designs, add nonlinearities across cells (which InceptionV3 does not implement by default), or increase or decrease how many cells you stack to adjust the network depth. Moreover, cell-based structures are incredibly simple and quick to implement, so this comes at little cost.

Neural Network Scaling

Successful neural network architectures are generally not built to be static in their size. Through some mechanism, these network architectures are *scalable* to different sets of problems. In the previous chapter, for instance, we explored how NASNet-style Neural Architecture Search design allowed for the development of successful cell architectures that could be scaled by stacking different lengths and combinations of the discovered cells. Indeed, a large advantage to cell-based design is inherent scalability. In this section, we'll discuss scaling principles that are applicable to all sorts of architectures – both cell-based and not.

The fundamental idea of scaling is that a network's "character" can be retained while the actual size of a network is scaled to be smaller or larger (Figure 6-42). Think of RV model airplanes – flyable airplanes that are only a few feet large in any dimension. They capture the spirit of what an airplane is by retaining its design and general function but use fewer resources by decreasing the size of each component. Of course, because they use fewer resources, they are less equipped for certain situations that true airplanes could withstand, like heavy gusts of wind, but this is a necessary sacrifice for scaling.

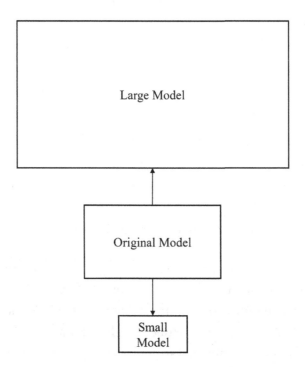

Figure 6-42. *Scaling a model to different sizes*

There are generally two dimensions that can be scaled – the width (i.e., the number of nodes, filters, etc. in each layer) and length (i.e., the number of layers) in the network. These two dimensions are easily defined in linear topologies but are more ambiguous in nonlinear topologies. To make scaling simple, there are generally two paths to deal with nonlinearity. If the nonlinearity is simple enough that there is an easily identifiable linear backbone (e.g., residual connection-based architectures or simple branch architectures), the linear backbone itself is scaled. If the nonlinearity is too complex, arrange it into cells that can be linearly stacked together and scaled depth-wise. We'll discuss these ideas more in detail.

Say, you have a neural network architecture on your hands – perhaps you've gotten it from a model repository or designed one yourself. Let's consider three general scenarios in which scaling can help:

- You find that it yields insufficiently high performance on your dataset, even when trained until fruition. You suspect that the maximum predictive capacity of the model is not enough to model your dataset.

- You find that the model performs fine as is, but you want to decrease its size to make it more portable in a systematic way without damaging the key attributes that contribute to the model's performance.

- You find that the model performs well as is, but you want to open-source the model for usage by the community or would like to employ it in some other context in which it will be applied to a diverse array of problem domains.

The principles of neural network scaling provide the key concepts to increase or decrease the size of your neural network in a systematic and structured fashion. After you have a candidate model architecture built, incorporating scalability into its design makes it more accessible across all sorts of problems.

Input Shape Adaptable Design

One method of scaling is to adapt the neural network to the input shape size. This is a direct method of scaling the architecture in relation to the complexity of the dataset, in which the relative resolution of the input shape is used to approximate the level of predictive power that needs to be allocated to process it adequately. Of course, it is not always true that the size of the input shape is indicative of the dataset's complexity to model. The primary idea of input shape adaptable design is that a larger input shape indicates more complexity and therefore requires more predictive power *relative* to a smaller input (the opposite relationship being true as well). Thus, a successful input shape adaptable design is able to modify the network in accordance to chances in the input shape.

This sort of scaling is a practical one, given that input shape is a key vital component of using and transferring model architectures (i.e., if it is not correctly set up, the code will not run). It allows you to directly build model architectures that respond to different resolutions, sizes, vocabulary lengths, etc. with correspondingly more or less predictive resource allocation.

The simplest adaptation to the input shape is not to modify the architecture at all, but to modify the shape of the input via a resizing layer (Listing 6-29). In Keras syntax, a None in a dimension of the input shape parameter indicates that the network accepts an arbitrary value for that dimension.

Listing 6-29. Adding a resizing layer with an adaptable input layer

```
inp = L.Input((None,None,3))
resize = L.experimental.preprocessing.Resizing(
    height=128, width=128
)(inp)
dense1 = L.Dense(64)(resize)
...
```

This sort of resizing design has the benefit of *portability* in deployment; it's easy to make valid predictions by passing in an image (or some other data form with appropriate code modifications) of any size into the network without any preprocessing. However, it doesn't quite count as *scaling* in that computational resources are not being allocated to adapt to the input shape. If we pass an image with a very high resolution – (1024, 1024, 3), for instance – into the network, it will lose a tremendous level of information by naïvely resizing to a certain height or width.

Let's consider a simple example of a fully connected network whose depth is fixed, but whose widths are open to be adjusted. It hypothetically contains layers that transform the input width from 32 to 21 to 14 to 10 to 10 to 5 to 1 (output) across five layers. We need to identify a scalable pattern – an *architectural policy* that can be generalized. In this case, each width is approximately two-thirds the width of the previous width. We can implement this recursive architectural policy by defining five layer widths based on this generalized policy and storing it in a list of layer widths (Listing 6-30).

Listing 6-30. Creating a list of widths via a recursive architectural policy

```
inp_width = 64
num_layers = 5

widths = [inp_width]
next_width = lambda w:round(w*2/3)
for i in range(num_layers):
    widths.append(next_width(widths[-1]))
```

We can apply this to the construction of our simple neural network (Listing 6-31).

Listing 6-31. Building a model with the scalable architectural policy to determine network widths. We begin reading from the i+1th index because the first index contains the input width

```
model = keras.models.Sequential()
model.add(L.Input((inp_width,)))
for i in range(num_layers):
    model.add(L.Dense(widths[i+1]))
model.add(L.Dense(1, activation='sigmoid'))
```

This model, albeit simple, is now capable of being transferred to datasets of different input sizes.

We can similarly apply these ideas to convolutional neural networks. Note that convolutional layers can accept arbitrarily sized inputs, but nevertheless we can change the number of filters used in each layer by finding and applying generalized rules in the existing architecture to be scaled. In convolutional neural networks, we want to expand the number of filters to compensate for decreases in image resolution. Thus, the number of filters should increase over time relative to the original input shape to perform this resource compensation properly.

We can define the number of filters we want the first convolutional layer and the last convolutional layer to have as dependent on the resolution of the input shape (Listing 6-32). In this case, we define it the number of filters as $2^{round(log_2 w)-4}$, where w is the original width (the expression is purposely left in unsimplified form). Thus, a 128x128 image would begin with 8 filters, whereas a 256x256 image would begin with 16 filters. The number of filters a network has is scaled up or down relative to the size of the input image. We define the number of filters the last convolutional layer has simply to be $2^3 = 8$ times the number of original filters.

Listing 6-32. Setting important parameters for convolutional scalable models

```
inp_shape = (128,128,3)
num_layers = 10

w, h, d = inp_shape
start_num_filters = 2**(round(np.log2(w))-4)
end_num_filters = 8*start_num_filters
```

Our goal is to progress from the starting number of filters to the ending number of filters throughout the ten layers we define our network depth to be. In order to do this, we divide the number of layers into four segments. To go from one segment to the next, we multiply the number of filters by 2. We determine when to move to the next segment by measuring what fraction of the layers have been accounted for (Listing 6-33).

Listing 6-33. Architectural policy to determine the number of filters at each convolutional layer

```
filters = []
for i in range(num_layers):
    progress = i/num_layers
    if progress < 1/4:
        f = start_num_filters
    elif progress < 2/4:
        f = start_num_filters*2
    elif progress < 3/4:
        f = start_num_filters*4
```

```
else:
    f = start_num_filters*8
filters.append(f)
```

The sequence of filters for a (128,128,3) image, according to this method, is [8, 8, 8, 16, 16, 32, 32, 32, 64, 64]. The sequence of filters for a (256,256,3) image is [16, 16, 16, 32, 32, 64, 64, 64, 128, 128]. Note that we don't define the architectural policy in this case recursively.

By determining which segment to enter by measuring *progress* rather than the deterministic layer (i.e., if the current layer is past Layer 3), this script is also scalable across *depth*. By adjusting the num_layers parameter, you can "shrink" or "stretch" this sequence across any desired depth. Because higher resolutions generally warrant longer depths, you can also generalize the num_layers parameter as a function of the input resolution, like num_layers = round(np.log2(128)*3). Note that in this case we are using the logarithm to prevent the scaling algorithm from constructing networks with depths that are too high in response to high-resolution inputs. The list and length of filters can then be used to automatically construct a neural network architecture scaled appropriately depending on the image.

Note that these operations should be fitted *after* you have obtained a model to scale in the first place. We can generalize this process of scaling into three steps:

1. Identify architectural patterns that can be generalizable for scaling, like the depth of the network or how the width of layers changes across the length of the network.

2. Generalize the architecture pattern into an *architectural policy* that is scalable across the scaled dimension.

3. Use the architectural policy to generate concrete elements of the network architecture in response to the input shape (or other determinants of scaling).

Adaptation to input shapes is most relevant for architectures like autoencoders, in which the input shape is a key influence throughout the architecture. We'll combine cell-based design and knowledge of autoencoders discussed in Chapter 3 to create a *scalable* autoencoder whose length and filter sizes can automatically be scaled to whatever input size it is being applied to.

To make the process of building easier, let's define an "encoder cell" and a "decoder cell" (Listing 6-34). These are not the encoder and decoder *sub-models* or components but rather represent cells that can be stacked together to form the encoder and decoder. The encoder cell attaches two convolutional layers and a max pooling layer to whatever layer is passed into the cell-building function, whereas the decoder cell attaches the "inverse" – an upsampling layer followed by two transpose convolutional layers. Both return the output/last layer of the cell, which can be used as the input to the following cell. Note that these two cells halve and double the resolution of image inputs, respectively.

Listing 6-34. Encoder and decoder cells, for example, autoencoder-like structure

```
def encoder_cell(inp_layer, filters):
    x = L.Conv2D(filters, (3,3), padding='same')(inp_layer)
    x = L.Conv2D(filters, (3,3), padding='same')(x)
    x = L.MaxPooling2D((2,2))(x)
    return x

def decoder_cell(inp_layer, filters):
    x = L.UpSampling2D((2,2))(inp_layer)
    x = L.Conv2DTranspose(filters, (3,3), padding='same')(x)
    x = L.Conv2DTranspose(filters, (3,3), padding='same')(x)
    return x
```

We will begin with three key variables: i will represent the power with which 2 is raised to in determining how many filters the first and last convolutional layer should have ($i=4$ indicates that $2^4 = 16$ filters will be used); w, h, and d will be used to hold the width, height, and depth of the input shape; and curr_shape will be used to track the shape of data as it is passed through the network (Listing 6-35).

Listing 6-35. Defining key factors in building the scalable autoencoder-like architecture

```
i = 4
w, h, d = (256,256,3)
curr_shape = np.array([w, h])
```

We must consider the two key dimensions of scale: width and depth.

The number of filters in a block will be doubled if the resolution is halved, and it will be halved if the resolution is doubled. This sort of relationship ensures approximate *representation equilibrium* throughout the network – we do not create representation bottlenecks that are too severe, a principle of network design outlined by Szegedy et al. in the Inception architecture (see case study for section on cell-based design). Assume that this network is used for a purpose like pretraining or denoising in which bottlenecks can be built less liberally in size (in other contexts, we would *want* to build more severe representation bottlenecks). We can keep track of this relationship by increasing i by 1 after an encoder cell is attached and decreasing it by 1 after a decoder cell is attached.

In this example, we will build our neural network not with a certain pre-specified depth, but with whatever depth is necessary to obtain a certain bottleneck size. That is, we will continue stacking encoder blocks until the data shape is equal to (or falls below) a certain desired bottleneck size. At this point, we will stack *decoder* blocks to progressively increase the size of the data until it reaches its original size.

We can construct the input layer and the first encoder cell (Listing 6-36). We update the current shape of the data appropriately by halving after an encoder cell is added. Moreover, we increase i such that the following encoder cell uses two times as many filters. An infinite loop continues stacking encoder cells until the output shape of the cell (i.e., the potential bottleneck) is equal to or less than 16 neurons (the desired cell).

Listing 6-36. Building the encoder component of the scalable autoencoder-like architecture

```
inp = L.Input((w,h,d))

x = encoder_cell(inp, 2**i)
curr_shape = curr_shape/2
i += 1

# build encoder
while True:
    x = encoder_cell(x, 2**i)
    curr_shape = curr_shape/2
    if curr_shape[0] <= 16: break
    i += 1
```

After the encoder is built, we can correspondingly build the decoder, which repeatedly stacks decoder cells and decreases i such that the following cell uses half as many filters (Listing 6-37). While you could continue to keep track of the shape and break when the current shape was equal to the original shape, another approach that uses less code is to take advantage of our usage of i and treating i=4 as an indication that we have reached the initial state.

Listing 6-37. Building the decoder component of the scalable autoencoder-like architecture

```
# build decoder
while True:
    x = decoder_cell(x, 2**i)
    if i == 4: break
    i -= 1
```

The complete model can be aggregated as ae = keras.models.Model(inputs=inp, outputs=x). This simple autoencoder design – using two convolutions followed by a pooling operation – has been scaled to be capable of modeling any input size resolution (as long as it is a power of 2, since pooling makes approximations that are not captured in upsampling if the side length is not cleanly divisible by the pooling factor – see Chapter 3 for more on this). Scaling a model to be adaptable to different input sizes can require more work, as we've seen, but it makes your model more accessible and agile in experimentation and deployment.

Parametrization of Network Dimensions

In the earlier section, we focused on scaling oriented toward a necessity-level parameter: the input shape of a model. In this section, we will discuss broad parametrization of network dimensions. Adapting the network architecture to the input shape requires us to generalize the model architecture and thus formulate implicit and explicit architectural policies that may or may not be successful. The goal here is rather to *parametrize* the dimensions of the network for the purposes both of adaptation to different problems and also for *experimentation.*

As discussed in the introduction to this chapter, seldom will one round of model building suffice for deployment. By parameterizing the dimensions of a network architecture, we are able to experiment with different scales and sizes more easily and quickly for a network to optimally fit a dataset.

The key difference between parametrizing a model for the sake of experimentation and inherent scalability and parametrizing a model to adapt to the input shape is that the factors determining parametrization are user-specified, not dependent on the input shape. Rather than programming architectural policies (e.g., the pattern with which the width of a network expands from the input shape), we use *multiplying coefficients*. These are parameters that the user specifies that are multiplied to the current dimensions of the network. A multiplying coefficient smaller than 1 will shrink that dimension, whereas a multiplying coefficient larger than 1 will expand that dimension.

Consider this simple sequential model architecture, which processes a 64-dimensional input through four fully connected layers and an output (Listing 6-38).

Listing 6-38. Building a simple sequential model to be parametrized

```
model = keras.models.Sequential()
model.add(L.Input(64,))
model.add(L.Dense(32, activation='relu'))
model.add(L.Dense(32, activation='relu'))
model.add(L.BatchNormalization())
model.add(L.Dense(16, activation='relu'))
model.add(L.Dense(16, activation='relu'))
model.add(L.Dense(1, activation='sigmoid'))
```

Let's parametrize the width by multiplying the number of nodes in each layer by some width coefficient (Listing 6-39). Because the result may be a fraction, we round the result of the scaling.

Listing 6-39. Parametrizing the width of a network

```
width_coef = 1.0
w = lambda width: round(width*width_coef)

model = keras.models.Sequential()
model.add(L.Input(64,))
model.add(L.Dense(w(32), activation='relu'))
```

```
model.add(L.Dense(w(32), activation='relu'))
model.add(L.BatchNormalization())
model.add(L.Dense(w(16), activation='relu'))
model.add(L.Dense(w(16), activation='relu'))
model.add(L.Dense(1, activation='sigmoid'))
```

Parametrizing the depth is a little bit more tricky, because we need to manipulate actual layer objects, rather than parameters within a fixed set of layers. A successful systematic approach is to identify key blocks of the architecture consisting of multiple similar layers that can be stretched or shrunk by a depth coefficient (Listing 6-40). In our simple model, there are two easily identifiable blocks: one block directly after the input and before the batch normalization consisting of two layers with 32 nodes and another block after the batch normalization layer consisting of another two layers with 16 nodes. By default, these two blocks consist of one type of layer with a default quantity of 2. We can parametrize the network, therefore, by multiplying this quantity by the depth coefficient. Like with the width, we perform rounding in the case of noninteger results.

Listing 6-40. Parametrizing the depth of a network

```
depth_coef = 1.0
d = lambda depth: round(depth*depth_coef)

model = keras.models.Sequential()
model.add(L.Input(64,))
for i in range(d(2)):
    model.add(L.Dense(w(32), activation='relu'))
model.add(L.BatchNormalization())
for i in range(d(2)):
    model.add(L.Dense(w(16), activation='relu'))
model.add(L.Dense(1, activation='sigmoid'))
```

In this case, we pass 2 into d() because 2 is the default number of layers in our model architecture. Note additionally that we are not scaling the *entire* depth of the network; we are leaving layers like batch normalization alone regardless of the depth coefficient, for instance. Depth-wise scaling should be applied appropriately to processing layers, not to layers like batch normalization or dropout that only shift or regularize the data flow.

The user can now adjust `width_coef` and `depth_coef` for quick experimentation and portability. The method by which you *optimize* the parametrization of network dimensions is up to you. One method that is likely to be successful is to use Bayesian optimization via Hyperopt or Hyperas to tune the width and depth scaling factors `width_coef` and `depth_coef`. Alternatively, one can look toward a recently growing body of research around general best practices for scaling, like the *compound scaling method* introduced in the successful EfficientNet architecture of models. We'll explore this method in the case study for this section.

This logic applies to architectures with nonlinearity, as long as a clear backbone is identifiable (Listing 6-41). Consider, for instance, the code used for building DenseNet-style residual connections (Listing 6-8). We can scale network depth and width from their original "default" dimension values using our d and w functions, for instance.

Listing 6-41. Parametrizing nonlinear architectures (in this case, DenseNet-style model) by relying upon a linear backbone. Complete code is not shown. Please refer to relevant DenseNet-style residual connection listing for full context

```
num_layers = d(5)

inp = L.Input((128,))
x = L.Dense(64, activation='relu')(inp)
layers = [x]
for i in range(num_layers-1):
    x = L.Dense(w(64))(make_rc(layers))
    layers.append(x)
output = L.Dense(1, activation='sigmoid')(x)
```

Similarly, you can parametrize the dimensions of nonlinear architectures without a nonlinear backbone that are simple, like parallel branches. If an architecture is too nonlinear to use a block-based approach toward scaling the depth dimension as introduced earlier, another method is to group these highly nonlinear topologies into blocks that can be scaled by stacking different quantities of blocks together.

By parameterizing network dimensions, you enable yourself and others to experiment and adapt the network architecture more quickly and easily, leading to improved performance on the problem.

Case Study: EfficientNet

Convolutional neural networks have historically been scaled relatively arbitrarily, along the two previously discussed dimensions – height and width – as well as (recently) resolution. "Arbitrary" scaling entails adjusting these dimensions of a network without much of a justification for how the adjusting is performed; there is ambiguity in how large to scale dimensions. The "larger is better" paradigm that gripped much of earlier development in convolutional neural network designs is reaching a limit in its competitiveness against other approaches that focus more on developing efficient mechanisms and designs. Thus, there is a need for a *systematic method of scaling neural network architectures across several dimensions* for the highest expected success (Figure 6-43).

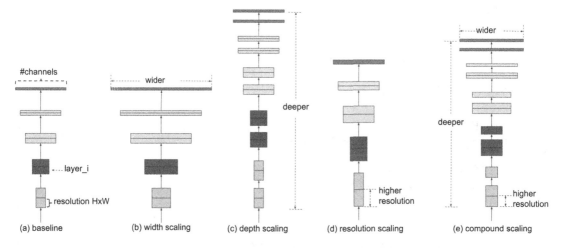

Figure 6-43. *Dimensions of a neural network that can be scaled, compared with the compound scaling method*

Mingxing Tan and Quoc V. Le propose the *compound scaling method* in their paper "EfficientNet: Rethinking Model Scaling for Convolutional Neural Networks."[3] The compound scaling method is a simple but successful scaling method in which each dimension is scaled by a constant ratio.

---

[3] Mingxing Tan and Quoc V. Le, "Rethinking Model Scaling for Convolutional Neural Networks," 2019. Paper link: https://arxiv.org/pdf/1905.11946.pdf.

A set of fixed scaling constants is used to uniform scale the width, depth, and resolution used by a neural network architecture. These constants – α, β, γ – are scaled by a compound coefficient, ϕ, such that the depth is $d = \alpha^{\phi}$, the width is $w = \beta^{\phi}$, and the resolution is $r = \gamma^{\phi}$. ϕ is defined by the user, depending on how many computational resources/predictive power they are willing to allocate toward a particular problem.

The values of the constants can be found through a small grid search, in which ϕ is set to 1 and the set of parameters that yield the best accuracy is selected. This is both feasible and successful given the small search space. Two constraints on the constants are imposed:

- $\alpha \geq 1, \beta \geq 1, \gamma \geq 1$: This ensures that the constants do not decrease in value when they are raised to the power of the compound coefficient ϕ, such that a larger compound coefficient value yields in a larger depth, width, and resolution size.

- $\alpha \cdot \beta^2 \cdot \gamma^2 \approx 2$: The FLOPS (floating point operations per second) of a series of convolution operations is proportional to the depth, the width squared, and the resolution squared. This is because depth operates linearly by stacking more layers, whereas the width and the resolution act upon two dimensional filter representations. To ensure computational interpretability, this constraint ensures that any value of ϕ will raise the total number of FLOPS by approximately $(\alpha \cdot \beta^2 \cdot \gamma^2)^{\phi} = 2^{\phi}$.

Using this method of scaling is very successful in application to previously successful architectures like MobileNet and ResNet (Table 6-4). Through the compound scaling method, we are able to expand the size and computational power of the network in a structured and non-arbitrary way that optimizes the resulting performance of the scaled model.

Table 6-4. *Performance of the compound scaling method on MobileNetV1,*
MobileNetV2, and ResNet50 architectures

Model	FLOPS	Top 1 Acc.
Baseline MobileNetV1	0.6 B	70.6%
Scale MobileNetV1 by width ($w = 2$)	2.2 B	74.2%
Scale MobileNetV1 by resolution ($r = 2$)	2.2 B	74.2%
Scale MobileNetV1 by compound scaling	**2.3 B**	**75.6%**
Baseline MobileNetV2	0.3 B	72.0%
Scale MobileNetV2 by depth ($d = 4$)	1.2 B	76.8%
Scale MobileNetV2 by width ($w = 2$)	1.1 B	76.4%
Scale MobileNetV2 by resolution ($r = 2$)	1.2 B	74.8%
Scale MobileNetV2 by compound scaling	**1.3 B**	**77.4%**
Baseline ResNet50	4.1 B	76.0%
Scale ResNet50 by depth ($d = 4$)	16.2 B	76.0%
Scale ResNet50 by width ($w = 2$)	14.7 B	77.7%
Scale ResNet50 by resolution ($r = 2$)	16.4 B	77.5%
Scale ResNet50 by compound scaling	**16.7 B**	**78.8%**

Tan and Le propose explanations for the success of compound scaling that are similar to our previous intuition developed when adapting the architecture of a neural network based on the input size. By intuition, when the input image is larger, all dimensions – not just one – need to be correspondingly increased to accommodate the increase in information. Greater depth is required to process the increased layers of complexity, and greater width is needed to capture the greater quantity of information. Tan and Le's work is novel in expressing the relationship between the network dimensions quantitatively.

Tan and Le's paper proposes the *EfficientNet* family of models, which is a family of differently sized models built from the compound scaling method. There are eight models in the EfficientNet family – EfficientNetB0, EfficientNetB1, ..., to EfficientNetB7, ordered from smallest to largest. The EfficientNetB0 architecture was discovered via Neural Architecture Search. In order to ensure that the derived model optimized both performance and FLOPS, the objective of the search was not merely to maximize the accuracy but to maximize a combination of performance and the FLOPS. The resulting architecture is then scaled using different values of ϕ to form the other seven EfficientNet models.

Note   The actual open-sourced EfficientNet models are slightly adapted from their pure scaled versions. As you may imagine, compound scaling is a successful but still approximate method, as is to be expected with most scaling techniques – these are generalizations across ranges of architecture sizes. To truly maximize performance, some fine-tuning of the architecture is still needed afterward. The publicly available versions of the EfficientNet model family contain some additional architectural changes after scaling via compound scaling to further improve performance.

The EfficientNet family of models impressively obtains higher performance on benchmark datasets like ImageNet, CIFAR-100, Flowers, and others than similarly sized models – both manually designed and NAS-discovered architectures (Figure 6-44). While the core EfficientNetB0 model was created as a product of Neural Architecture Search, the remaining members of the EfficientNet family are constructed via scaling.

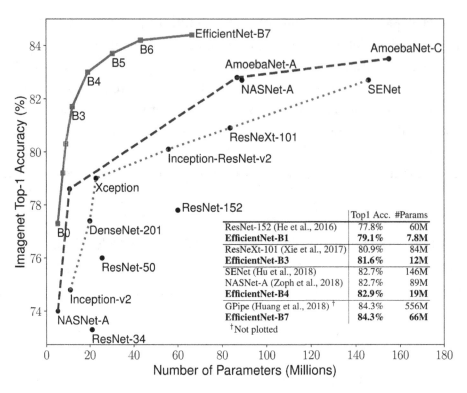

Figure 6-44. *Plot of various EfficientNet models against other important model architectures in number of parameters and ImageNet top 1 accuracy*

The EfficientNet model family is available in Keras applications at `keras.applications.EfficientNetBx` (substituting x for any number from 0 to 7). The EfficientNet implementation in Keras applications ranges from 29 MB (B0) to 256 MB (B7) in size and from 5,330,571 parameters (B0) to 66,658,687 parameters (B7). Note that the input shapes for different members of the EfficientNet family are different. EfficientNetB0 expects images of spatial dimension (224, 224); B4 expects (380, 380); B7 expects (600, 600). You can find this information in the expected input shape listed in Keras/TensorFlow applications documentation.

Looking at the EfficientNet source code in Keras is a valuable way to get a feel for how scaling is implemented on a professional level. It is available at `https://github.com/keras-team/keras-applications/blob/master/keras_applications/efficientnet.py`. Because this implementation is written for one of the most widely used deep learning libraries, much of the relevant code is used to generalize/parametrize the model for accessibility across various platforms and purposes. Nevertheless, its general structure of organization can be emulated for your deep learning purposes and designs.

The EfficientNet implementation in Keras consists of three key types of functions:

- block(), which builds a standard EfficientNet-style block given a long list of parameters, including the dropout rate, the number of incoming filters, the number of outgoing filters, etc. expand_ratio and se_ratio parameters refer to the "severity" or "intensity" of the *expansion phase* and the *squeeze and excitation phase*, which (roughly speaking) increase and decrease the representation size of the data.

- EfficientNet(), which constructs an EfficientNet model given two key parameters – the width_coefficient and the depth_coefficient. Additional parameters include the dropout rate and the depth divisor (a unit used for the width of a network).

- EfficientNetBx(), which simply calls the EfficientNet() architecture construction function with certain parameters appropriate to the EfficientNet structure being called. For instance, the EfficientNetB4() function returns the EfficientNet() function with a width coefficient of 1.4 and a depth coefficient of 1.8. The "unscaled" EfficientNetB0 model uses a width coefficient and depth coefficient of 1.

The key EfficientNet() function defines two functions within itself, round_filters and round_repeats, which take in the "default" number of filters and number of repeats and scale them appropriately depending on the provided width coefficient and depth coefficient.

The round_filters function (Listing 6-42) takes in the default number of filters and returns the new number of filters after scaling. The equation for the new number of

filters f^n is given roughly by $f_n = max\left(d, round\left(\dfrac{w \times f_o + \dfrac{d}{2}}{d}\right) \cdot d\right)$, where d is the depth

divisor, w is the width scaling coefficient, and fo is the original number of filters. The number of new filters can never be shrunk below the value of the divisor because of the *max*(…) mechanism. The expression on the right simply multiplies the original number of filters by the width scaling coefficient and then applies the *depth divisor* to it.

The depth divisor can be thought of as the bin size in quantization from Chapter 5; it is the basic unit that a parameter is scaled in terms of. The default divisor for EfficientNet is 8, meaning that the width is represented in multiples of 8. This sort of "quantization" can be easily done via the $round\left(\dfrac{a}{d}\right) \cdot d$ (integer division is performed in Python by a//d). This implementation adds $\dfrac{d}{2}$ to "balance" the scaled number of filters before "quantization"/"rounding."

Listing 6-42. Keras EfficientNet implementation of the function used to return the scaled width of a layer

```
def round_filters(filters, divisor=depth_divisor):
    filters *= width_coefficient
    new_filters = max(divisor, int(filters + divisor / 2)
    // divisor * divisor)
    if new_filters < 0.9 * filters:
        new_filters += divisor
    return int(new_filters)
```

The depth scaling method is simpler; the default number of block repeats is multiplied by the depth coefficient, with the ceiling function applied for a resulting noninteger scaled depth (Listing 6-43).

Listing 6-43. Keras EfficientNet implementation of the function used to return the scaled depth of the network

```
def round_repeats(repeats):
    return int(math.ceil(depth_coefficient * repeats))
```

These functions are used in the building of the parametrized EfficientNet base model, allowing for easy scaling.

Key Points

In this chapter, we discussed three key themes in successful neural network architecture design: nonlinear and parallel representation, cell-based design, and architecture scaling:

- There are three key concepts in efficient and advanced implementation of complex architectures – compartmentalization, automation, and parametrization.

- Nonlinear and parallel representations allow layers to pass information signals across various components of the architecture without being restricted by having to pass through many other components. This allows for the network to process information in a way that considers more perspectives and representations.

  - Residual connections are connections that "skip" or "jump" over other layers. ResNet-style residual connections are used repeatedly to jump over small stacks of layers. DenseNet-style residual connections, on the other hand, place residual connections between every pair of anchor points, allowing information to traverse both longer and shorter distances through the network. Residual connections are one method of addressing the vanishing gradient problem. These can be implemented quite simply through the Functional API by merging the "root" of the residual connection with the input to the "end" of the residual connection.

  - Branching structures and cardinality are generalizations of residual connections into broader nonlinearities. While width measures how wide one layer is (e.g., number of nodes or filters), *cardinality* measures how many layers wide – and therefore how many parallel representations exist and are being processed – the network is at some point.

- Blocks/cell design consists of arranging layers into packaged topologies that function as cells which can be stacked upon one another to form a cell-based architecture. By arranging layers into cells and manipulating cells rather than layers, we replace the base unit of architectural construction – the layer – with a more powerful one, consisting of an agglomeration of layers. Cells can be thought of as "mini-networks" that can take on a variety of internal topologies, linear or nonlinear. In implementation, block/cell design can be implemented by constructing a function which takes in a layer to build the cell on and outputs the last layer of the cell (upon which another cell or other processing layers can be stacked).

- Neural network scaling allows network architectures to be scaled for different datasets, problems, and experimentation. You can use scaling to adapt the architecture width and dimension to the input shape by identifying patterns in the neural network architecture and generalizing into an *architectural policy*. You can also scale architectures by parameterizing the width and the depth; these can be optimized via a method like Bayesian optimization or by a manual scaling policy like compound scaling.

In the next chapter, we will use the tools we've built across these multiple chapters to discuss deep learning problem-solving methods.

CHAPTER 7

Reframing Difficult Deep Learning Problems

Successful transformations reframe the problem that makes the solution possible. They erase existing boundaries and start from scratch.

—Malcolm Gladwell, Journalist and Author

To speak very roughly, the deep learning process can be represented with three central sequentially connected components: the creation of the architecture, the definition of training procedures and methods, and the training of the model itself. This is the general deep learning modeling workflow. Each of the chapters this book has covered up to this point can be categorized approximately into at least one of these categories. Chapter 2, on transfer learning and pretraining, discussed methods of training models to transfer and develop knowledge from sources other than the original dataset. Chapter 3, on autoencoders, discussed various usages of the versatile autoencoder concept and architecture. Chapter 4, on model compression, discussed various modifications to the neural network architecture, in addition to alterations in the training procedure. Chapter 5, on meta-optimization, discussed the automation of parameters in neural network architecture and training procedures. Chapter 6, on successful neural network design, discussed successful design patterns and techniques in neural network architectures and model implementation.

© Andre Ye 2022
A. Ye, *Modern Deep Learning Design and Application Development*,
https://doi.org/10.1007/978-1-4842-7413-2_7

In this last chapter – Chapter 7 – we will be taking a step back, moving away from strictly the study of building successful and efficient neural network systems and toward a more broad perspective on *using*, not just *doing*, deep learning in real-world problems. In order to do this, we're going to need to be comfortable with *reframing* difficult real-world problems that often don't come in a format that is convenient for successful application of deep learning. Although deep learning is an incredibly powerful and versatile tool, it often requires the human knack for problem-solving in the face of difficult tasks.

The key theme we will explore in this chapter is that of *data*. Data is the basis of deep learning problems, and it is a key factor in informing how one approaches the key components of deep learning (Figure 7-1).

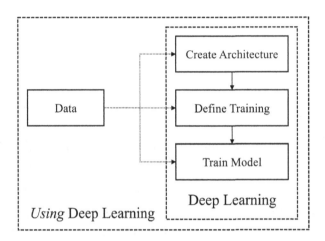

Figure 7-1. *Deep learning vs. deep learning: the role data plays in influencing each of the components of deep learning*

We've reached a point in which it's difficult to abstract the notion of "reframing the problem" into a generalized theory or a simple set of bullet points. This chapter will present three case studies as examples of successful reframing of difficult deep learning problems, in relation to data: experimentation with data representation, corrupted data, and limited data. While these case studies are not to be taken as strictly representative of the problems that you will face and need to reframe, they demonstrate that innovation comes with pushing and breaking boundaries. In doing so, hopefully they give you the creative spark to begin tackling new and unseen problems yourself.

Data Representation: DeepInsight

Data is a valuable commodity whose purity has become more and more prized throughout deep learning development. Whereas traditional machine learning methods often heavily relied upon human feature cleaning, engineering, and alteration, modern deep learning pushes toward a continual decreasing dependency on humans to perform data preprocessing by automating the preparation of the data, most commonly through clever neural network architecture designs.

Deep learning is automating the transgression of data forms; different methods of representing data are being bridged together in an automated fashion. Yet, there still remains a divide in data that deep learning still requires the guidance of humans to navigate – the *contextual dimension* of the data. By this, we mean not merely the dimensions (i.e., shape) of the data or the encoded knowledge it represents, but the organization of data along various dimensions of context from the data's origin, like spatial dimensions of width, height, and channels in an image, the dimension of time in audio inputs, and the dimension of sequence in text inputs. The contextual dimension of the data best describes what we understand to be meaningfully different "types" or "forms" of data in the context of deep learning.

So-called "specialized" data forms like images are generally reduced to the basic component of deep learning, the vector, upon which they can be processed as a primary element among standard layers and manipulations. This has been the general convention in the progression of deep learning development: the concept of feature extraction implies a tendency of complex data toward simplicity; the important elements of complex data are extracted, and from these elements, even more important elements are extracted, etc. Thus, complex data forms are reduced to the processing unit of the vector.

This is a rather unexplicit assumption or tradition in deep learning development, from which reframing the design paradigm and transgressing boundaries may yield new innovative and effective approaches. We can reframe the traditional flow of image data to vector form as vector data flowing into *image* form. Convolutional operations provide a local approach to feature extraction – that is, complex arrangements of data are progressively smoothed with local filters such that information is *progressively* generalized – that fully connected layers do not offer, viewing and processing all information at once. Intuitively, we may hypothesize that local extraction of features may prove to be valuable in domains other than just images.

Alok Sharma, Edwin Vans, Daichi Shigemizu, Keith A. Boroevich, and Tatsuhiko Tsunoda reframe this traditional design paradigm in their 2019 paper "DeepInsight: A methodology to transform a non-image data to an image for convolution neural network architecture."[1] The DeepInsight method is a pipeline to transform structured/ tabular data (this does not bar sequential or text-based data, as long as it is framed in a structured data format) into image-based data, which is then used for training in a standard convolutional neural network.

The first step of DeepInsight is to acquire a feature matrix that is used to map individual features in structured data to spatial coordinates in the corresponding image. This feature density matrix is used as a "template" to generate individual images for each individual vector. Each feature is associated with a pixel in the "template" matrix. This association is performed via a clever trick in which the data is *transposed* and used in dimensionality reduction via methods like Kernel PCA or t-Stochastic Neighbor Embedding. Traditionally, in a dataset of n samples and d features, dimensionality reduction to two spatial dimensions yields a dataset of n samples and two features. However, if we apply dimensionality reduction to the *transpose* of such a dataset, we treat each of the d features as a sample and each of the n samples as a feature, yielding a reduced dataset with d samples and two features. Thus, each of the d features has been mapped to a two-dimensional point in the "template" matrix.

Using a transformation method like Kernel PCA and t-SNE that preserves local relationships, we can map features that behave similarly to one another to physically closer locations in the feature matrix (Figure 7-2). This allows similar features in the generated images to be processed more efficiently by convolutions.

[1] Alok Sharma, Edwin Vans, Daichi Shigemizu, Keith A. Boroevich, and Tatsuhiko Tsunoda, "DeepInsight: A methodology to transform a non-image data to an image for convolution neural architecture," 2019. Paper link: www.nature.com/articles/s41598-019-47765-6.pdf.

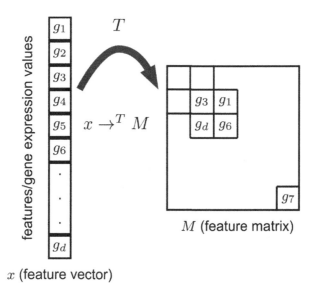

Figure 7-2. *Mapping features into locations in the template matrix m. g$_i$ is used to indicate a particular feature. $\to^T$ indicates the transformation procedure*

Once the "template" feature matrix has been established, we can create an image for an input vector by establishing a point in the image corresponding to the location allocated for that specific feature (Figure 7-3). (You can observe from this that DeepInsight is designed on high-dimensional data; a high number of features is needed to populate the image since each point in the image is one feature.) In order to prevent redundancy in image representation, the convex hull algorithm is utilized to select the smallest rectangle consisting of all the data, cropping out unnecessary blank sides. The data is correspondingly rotated, and the space is mapped into pixel, image-based format, which can then be passed through a standard convolutional neural network.

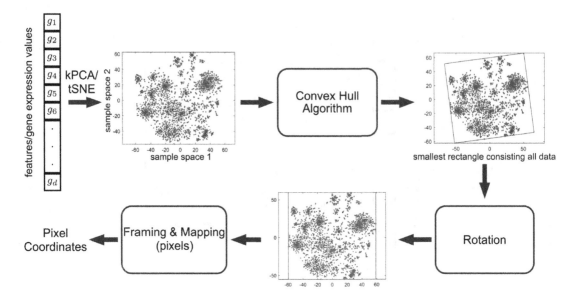

Figure 7-3. *The DeepInsight pipeline: mapping vectors to pixel coordinates*

Sharma et al. use a model to process this structured turned image data that combines many of the previously discussed in this book. The architecture (shown in Figure 7-4) used to process the images uses parallel representation with a cardinality of two, in which each of the branches is a simple linear stack of linear cells, which follow the standard convolutional pooling dynamic (with additional batch normalization and ReLU layers). The two branches are merged and processed for the classification output. The difference between the two branches lies in the filter size of image operations; by using an architecture that explicitly captures differently sized patterns within the image, the model is more effectively able to look for both broader patterns and smaller nuances simultaneously and combine the findings of these two perspectives to form a more accurate judgment. The number of cell repeats, filter sizes, learning rates, and other hyperparameters are tuned using Bayesian optimization.

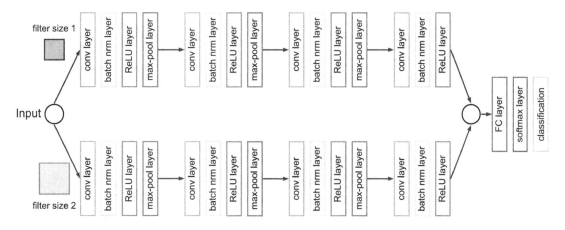

Figure 7-4. *Architecture used for DeepInsight pipeline*

The resulting DeepInsight pipeline performs very well on both genetic data, which the model was originally designed for, and other high-dimensional data contexts. Sharma et al. evaluate the method on five benchmark datasets: RNA-seq, a biological RNA sequence dataset from the NIH TCGA dataset; a subset from the TIMIT corpus, a speech dataset; the Relathe dataset, derived from news documents; the Madelon dataset, which is a synthetically constructed binary classification problem; and ringnorm DELVE, another synthetically constructed binary classification problem. These five datasets represent a wide array of problem contexts and data spaces; the DeepInsight method performs much better than other algorithms that have become successful staple methods in modeling structured/tabular datasets (Table 7-1). See Figure 7-5 for a visualization on how DeepInsight generates meaningful visual representations across these datasets.

Table 7-1. *Performance of DeepInsight against other common methods for structured data with various datasets*

Dataset	Decision Tree	AdaBoost	Random Forest	DeepInsight
RNA-seq	85%	84%	96%	99%
Vowels	75%	45%	90%	97%
Text	87%	85%	90%	92%
Madelon	65%	60%	62%	88%
Ringnorm DELVE	90%	93%	94%	98%

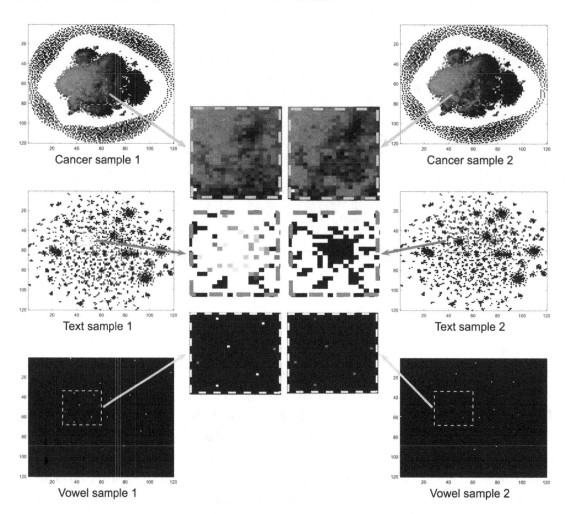

Figure 7-5. *Visualization of image patterns mapped by DeepInsight. Differences in samples from Cancer, Text, and Vowel datasets are shown as the small patches in the middle columns. These differences are extracted by convolutional filters to perform more effective classification than another method*

We can reflect on key advantages of DeepInsight and its subversion of the traditional image-to-vector pipeline with respect to processing structured data:

- CNNs do not require any additional feature extraction techniques, which are often used for structured data. They automatically derive advanced and information-rich features from the raw input data without need for preprocessing via a series of convolutions and pooling. The nonlinear architecture of the model used to process the image aids in the development of advanced, rich representations.

- Convolutions process image data locally in a restricted subarea. This allows for greater network depth with a relatively small quantity of parameters, which spurs healthy network understanding and generalization. Such performance would be more difficult if using a fully connected network, in which building more depth to increase modeling power leads to a faster increase in parameters, risking overfitting, which neural networks trained on structured data are especially prone to.

- The unique structure of the CNN allows it to run very efficiently, given recent hardware advancements like GPU utilization.

- CNNs and the DeepInsight pipeline broadly are much more customizable/optimizable than standard algorithms like tree-based methods that have traditionally shown success in modeling structured data. Besides adjusting hyperparameters like the model architecture, the vector-to-template matrix mapping, the learning rate, among countless others, one can also easily use image augmentation methods to generate "new" image data. Such data augmentation is difficult to accomplish with tabular data because the structured format does not contain an inherent robustness due to the low dimensionality of its representation space relative to images; that is, while rotating an image should not affect the phenomenon it represents, altering structured data likely will.

In practice, DeepInsight should be a contributing member in an ensemble of other decision-making models. Combining the locality-specific nature of DeepInsight with the more global approach of other modeling methods will likely yield a more informed prediction.

To implement DeepInsight, we'll be using code created by Sharma et al., which can be installed from the GitHub repository (Listing 7-1).

Listing 7-1. Installing code provided by Alok et al. for DeepInsight

```
!pip install git+git://github.com/alok-ai-lab/DeepInsight.git#egg=DeepInsight
```

The dataset we will use is the Mice Protein Expression Dataset from the infamous University of California Irvine Machine Learning Repository, which is a classification dataset with 1080 instances and 80 features modeling the expression of 77 proteins in the cerebral cortex of mice exposed to context fear conditioning. A cleaned version of the dataset is available in the source code for this book to be downloaded.

Assuming that the data has been loaded as a pandas DataFrame in the variable data, the first step is to separate into training and testing datasets, a standard procedure in machine learning (Listing 7-2). We'll also need to convert the labels to one-hot format, which in their original organization are integers corresponding to a class. This can be accomplished easily using keras.util's to_categorical function.

Listing 7-2. Selecting a subset of data and converting to one-hot form as necessary

```
import pandas as pd
# download csv from online source files
data = pd.read_csv('mouse-protein-expression.csv')
from sklearn.model_selection import train_test_split
X_train, X_test, y_train, y_test = train_test_split(data.drop('class',axis=1),
                    data['class'], train_size=0.8)
y_train = keras.utils.to_categorical(y_train)
y_test = keras.utils.to_categorical(y_test)
```

We will need to use the LogScaler object from the DeepInsight library to scale the data between 0 and 1 using the L2 norm (Listing 7-3). We transform both the training dataset and the testing dataset, fitting the scaler on the training dataset only. All new data used for prediction by the DeepInsight model should pass through this scaler first.

Listing 7-3. Scaling data

```
from pyDeepInsight import LogScaler
ln = LogScaler()
X_train_norm = ln.fit_transform(X_train)
X_test_norm = ln.transform(X_test)
```

The ImageTransformer object performs the image transformation by first generating the "template" matrix via a dimensionality reduction method passed into feature_extractor, which accepts either 'tsne', 'pca', or 'kpca'. This method is used to determine a mapping of features in the input vector to an image of pixels dimensions. We can instantiate an ImageTransformer with Kernel PCA dimensionality reduction method to generate 32x32 images (feature_extractor='kpca', pixels=32) (Listing 7-4).

Listing 7-4. Training and transforming with the ImageTransformer

```
from pyDeepInsight import ImageTransformer
it = ImageTransformer(feature_extractor='kpca', pixels=32)
tf_train_x = it.fit_transform(X_train_norm)
tf_test_x = it.transform(X_test_norm)
```

Kernel PCA is used rather than t-SNE because of the relatively low dimensionality and quantity of the data. PCA is not employed because its linearity limits the nuance it captures. An image length of 32 pixels is chosen as a balance between making the generated images too sparse (too high an image length) and too small (too small an image length) to meaningfully and accurately represent spatial relationships between features. As image size decreases, the conception of distance in the DeepInsight pipeline – that is, the placement of features as pixels farther or closer to one another dependent on their similarity – becomes more approximated to the point of being arbitrary.

We can visualize the generated images of the ImageTransformer easily using matplotlib.pyplot.imshow() to get a feel for how dimensionality reduction method and image size influence the arrangement of the features and the likelihood of success (Figure 7-6). The differences between images are subtle, but the distinguishing factors are identified and amplified by a series of convolutional operations. Note that the 32x32 pixel space allows for a clustering of similar features and for less related features to be distanced farther away in a corner.

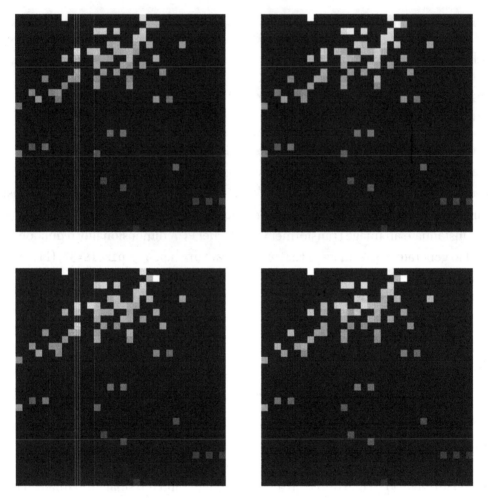

Figure 7-6. *Four example images generated from structured data using the DeepInsight method*

We'll build an architecture in a similar style to the two-branch cell design used in the DeepInsight paper, with three key adaptations: InceptionV3-style factorization/expansion of filters, dropout in cells, and a longer fully connected component (Listing 7-5). These aid to develop more specific filters with smaller areas to better parse densely packed features, further aid generalization by preventing overfitting, and better process derived features, respectively. One branch processes images with a kernel of size (2,2), and the other uses a kernel of size (5,5) (with additional factorization, e.g., 5x1 and 1x5).

Listing 7-5. Sample implemented architecture in Keras. You can, of course, optimize the architecture using the meta-optimization methods discussed in Chapter 5

```
# input
inp = L.Input((32,32,3))

# branch 1
x = inp
for i in range(3):
    x = L.Conv2D(2**(i+3), (2,1), padding='same')(x)
    x = L.Conv2D(2**(i+3), (1,2), padding='same')(x)
    x = L.Conv2D(2**(i+3), (2,2), padding='same')(x)
    x = L.BatchNormalization()(x)
    x = L.Activation('relu')(x)
    x = L.MaxPooling2D((2,2))(x)
    x = L.Dropout(0.3)(x)
x = L.Conv2D(64, (2,2), padding='same')(x)
x = L.BatchNormalization()(x)
branch_1 = L.Activation('relu')(x)

# branch 2
x = inp
for i in range(3):
    x = L.Conv2D(2**(i+3), (5,1), padding='same')(x)
    x = L.Conv2D(2**(i+3), (1,5), padding='same')(x)
    x = L.Conv2D(2**(i+3), (5,5), padding='same')(x)
    x = L.BatchNormalization()(x)
    x = L.Activation('relu')(x)
    x = L.MaxPooling2D((2,2))(x)
    x = L.Dropout(0.3)(x)
x = L.Conv2D(64, (5,5), padding='same')(x)
x = L.BatchNormalization()(x)
branch_2 = L.Activation('relu')(x)

# concatenate + output
concat = L.Concatenate()([branch_1, branch_2])
```

```
global_pool = L.GlobalAveragePooling2D()(concat)
fc1 = L.Dense(32, activation='relu')(global_pool)
fc2 = L.Dense(32, activation='relu')(fc1)
fc3 = L.Dense(32, activation='relu')(fc2)
out = L.Dense(9, activation='softmax')(fc3)

# aggregate into model
model = keras.models.Model(inputs=inp, outputs=out)
```

The model, when compiled and trained on the data for several dozen epochs, yields almost perfect training and accuracy *and* test accuracy (Listing 7-6).

Listing 7-6. Compiling and fitting the model

```
model.compile(optimizer='adam', loss='categorical_crossentropy',
              metrics=['accuracy'])
model.fit(tf_train_x, y_train, epochs=100, validation_data=(tf_test_x,y_test))
```

By expanding the data type with which convolutional and other image-based images can reach to structured data, DeepInsight bridges entire regions of deep learning and, in this case, has yielded an incredibly successful model.

Corrupted Data: Negative Learning with Noisy Labels

Improperly labeled data is more prevalent than we would like to think, especially in the context of annotated datasets, in which samples are manually labeled by human annotators that often move quickly and can make mistakes. Alternatively, adversarial attacks may be conducted on datasets that switch the labels of certain training items to maximally derail neural network performance. Since data that does not model the phenomena it represents faithfully yields in a model that does not properly model the phenomena it is trained to, often overfitting and developing "confused" representations (i.e., GIGO (Garbage In, Garbage Out)), there is a need for developing methods to handle data with corrupted labels that does not entail manually searching through the entire dataset to correct them.

Youngdong Kim, Junho Yim, Juseung Yun, and Junmo Kim propose a novel, simple but successful learning pipeline to deal with corrupted labels.[2] Their approach reframes the traditional approach to multiclass image classification tasks, *positive learning*, in which the neural network is "taught" to associate an image with a label – that is, this image *is* a [label]. *Negative learning*, on the other hand, is when the neural network is "taught" to *not* associate an image with a label – that is, this image is *not* a [label] (Figure 7-7).

Given noisy label : Car

Figure 7-7. *Visual difference between positive learning and negative learning*

In a multiclass context, negative learning is an indirect method of learning. Several negative learning associations must be learned to be equivalent to one positive learning association. For instance, if we are training a model to classify MNIST-style digits from 0 to 9, to represent the knowledge that some image represents the digit "3," we would need nine negative learning associations: that image is *not* the digit "0," that image is *not* the digit "1," …, that image is *not* the digit "9." The point of negative learning in this context, though, is not to equivalently represent positive learning because we know that many of the positive learning associations have been corrupted.

Suppose we are training a model on a multiclass problem with k classes. Rather than using positive learning by associating some training input x with a positive label y_p, which we know may be corrupted, we randomly select one of the $k-1$ other classes to use as a negative label y_n. If the original positive label *was* corrupted (i.e., y_p does not match x), then there is a $1-\dfrac{1}{k-1}$ probability that a randomly selected negative label is true – only one of the other $k-1$ negative labels is false when applied to x because it

[2] Youngdong Kim, Junho Yim, Juseung Yun, and Junmo Kim, "Negative Learning for Noisy Labels," 2019. Paper link: https://arxiv.org/pdf/1908.07387.pdf.

is the positive label. As k increases, the probability that a randomly selected negative label is true grows closer to 1. For modern datasets, with $k = 100$ or even $k = 1000$, the probability that a negative label is incorrect is negligible. On the other hand, if the original positive label is *not* corrupted (i.e., it is correct), then there is a 100% chance that a randomly selected alternative negative label is true.

Training a model on negative labels necessitates a different loss function. Suppose the true one-hot vector for a five-class classification problem is [1, 0, 0, 0, 0] (i.e., the negative label is the class index 0 represents) and the prediction is [0.3, 0.4, 0.2, 0.0, 0.1]. Our loss function does not need to consider predictions for classes other than the negative label. All we care about is that the model's prediction for the class at index 0 (the 0.3 prediction) decreases toward zero, because we want the model to be as least confident as possible with respect to that class. Rather than increasing the probability of the correct class, the model is trained to minimize the confidence of a randomly selected incorrect class (i.e., corresponding to the negative label).

Besides ensuring an increased truthfulness of the labels, this method of negative learning discourages overfitting to the noisy data by introducing associations via indirect associations. In a sufficiently large dataset, these indirect associations learned via negative learning compound together to form knowledge comparable to that of positive learning (if the labels weren't corrupted). Evidence of lack of overfitting is demonstrated by comparing the probability distribution of confidences for models trained on positive learning and negative learning – a model trained with negative learning is less confident and more accurately identifies corrupted examples (Figure 7-8).

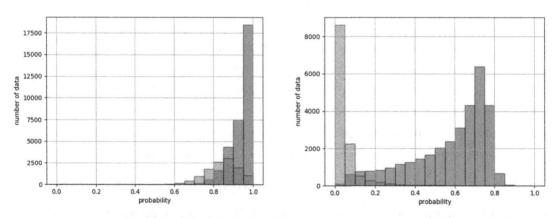

Figure 7-8. *Left: two probability distributions representing model confidence in datasets (left, orange: corrupted data; right, blue: uncorrupted data) with just positive learning. Right: with just negative learning*

To improve upon convergence, the authors introduce *selective negative learning* (SelNL), in which the convolutional neural network is trained further only on data with which the model predicts a label with a confidence larger than $\frac{1}{k}$ (larger than chance). This means that negative labels that may be incorrect contradict the learned knowledge of other correct negative labels and thus can be filtered out of the training dataset. SelNL helps the model both become more confident and more accurately separate the probability distributions of corrupted vs. uncorrupted data (Figure 7-9).

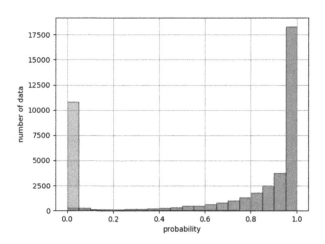

Figure 7-9. *Two probability distributions representing model confidence in datasets (left, orange: corrupted data; right, blue: uncorrupted data) with NL → SelNL*

Kim et al. introduce another training paradigm, *selective positive learning* (SelPL), on the basis that positive learning is faster and more accurate than negative learning given correct labels. Now that the negative learning and selective negative learning paradigm has allowed for a more or less accurate separation of corrupted and uncorrupted data, we can filter out data with which the model predicts with a confidence less than γ. Data that passes this threshold requirement are considered to be clean, and the *positive labels* for the clean data are used for further training on the network.

At the end of this pipeline – *NL → SelNL → SelPL*, termed the SelNLPL pipeline – the model is able to clearly separate corrupted and uncorrupted data, attaching a low probability to the former and a high probability to the latter (Figure 7-10).

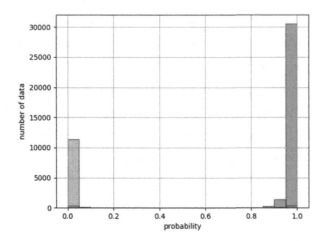

Figure 7-10. *Two probability distributions representing model confidence in datasets (left, orange: corrupted data; right, blue: uncorrupted data) with NL→ SelNL→ SelPL*

Note that the network's transition from negative learning to positive learning in the *SelNL → SelPL* phase may seem to be jarring but has been empirically observed to be smooth. The network needs only to rearrange and reformat the knowledge it has already obtained to acclimate to the positive learning regime. This is demonstrated by the training curve of the model as it is trained in various stages: negative learning, selective negative learning, and selective positive learning (Figure 7-11).

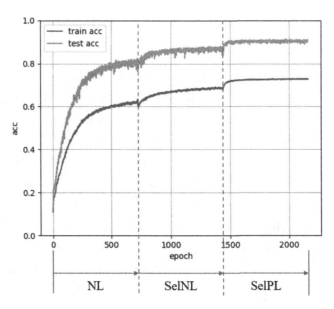

Figure 7-11. *Performance across the NL → SelNL → SelPL pipeline*

This model's filtering capabilities can be used not only to identify corrupted data but also to correct it. A new classification network is trained on the clean dataset (the set of data that has been identified by the previous model not to be corrupted). It then predicts the labels of the corrupted data to update their labels. The entire datasets – both the originally uncorrupted and the updated data – are used to train a final classification network. The final classification network has access to a corrected, almost completely truthful dataset (Figure 7-12). Thus, it can more faithfully model the phenomena the data represents.

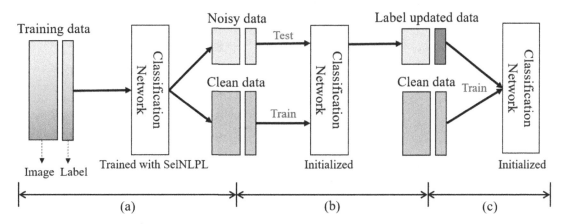

Figure 7-12. *Complete NLNL pipeline – SelNLPL pipeline to separation in noisy and clean data to label updating to final model classification*

This pipeline achieves state-of-the-art performance in the realm of addressing data corruption, performing better than other data corruption methods across a wide span of models and datasets (Table 7-2).

Table 7-2. *Performance of NLNL pipeline against other methods*

Dataset	Model	Method	Noise Level Performance		
			20%	40%	60%
FashionMNIST	ResNet18	CE	93.24	92.09	90.29
		MAE	80.39	79.30	82.41
		Truncated L_q	93.21	92.60	91.56
		NLNL	**94.82**	**94.16**	**92.78**
CIFAR-10	ResNet14	CE	83.7	–	–
		Bootstrap – soft	84.3	–	–
		Bootstrap – hard	83.6	–	–
		NLNL	**89.85**	–	–
	ResNet34	CE	86.98	81.88	74.14
		MAE	83.72	67.00	64.21
		Truncated L_q	89.70	87.62	82.70
		NLNL	**94.23**	**92.43**	**88.32**
CIFAR-100	ResNet34	CE	58.72	48.20	37.41
		MAE	15.80	9.03	7.74
		Truncated L_q	67.61	62.64	54.04
		NLNL	**71.52**	**66.39**	**56.51**
MNIST	LeNet	CE	88.02	68.46	45.51
		Boot – hard	87.69	69.49	50.45
		Boot – soft	88.50	70.19	46.04
		D2L	98.84	98.49	94.73
		NLNL	**99.35**	**99.27**	**98.91**

By reframing the traditional positive learning paradigm as negative learning and then using it to build a complete learning and filtering pipeline, the Negative Learning for Noisy Labels method cleverly solves the difficult problem of corrupted labels.

We will be implementing a slightly revised/simplified version of the NLNL pipeline proposed by Kim et al. We will use the CIFAR-10 dataset, which contains several thousand 32x32 pixel images from ten different classes of animals and transportation. This dataset can be loaded directly from Keras datasets (Listing 7-7).

Listing 7-7. Load CIFAR-10 dataset

```
(x_train, y_train), (x_test, y_test) = keras.datasets.cifar10.load_data()
```

First, we need to inject noise into the dataset (Listing 7-8). We will do this simply by randomly switching a certain subset of labels to another different label. A subset of indices are initially randomly selected via numpy's random choice function and will be used to determine which data instances will be corrupted.

Listing 7-8. Selecting random indices for corruption. We specify replace=False in the np.random.choice function to indicate that we are drawing from the list of indices, created by np.arange(len(x_train)), without replacement. This avoids random selection of duplicate indices

```
perc = 0.2
from np.random import choice
selected_indices = choice(np.arange(len(x_train)),
                          int(round(perc*len(x_train))), replace=False))
```

To perform the corruption (Listing 7-9), we can loop through each of the randomly selected indices. For each one, we generate a list of possible indices (all classes except for the true class) and randomly choose from this set of possibilities. These changes are made to cy_train, designated to store the corrupted *y* labels for training. This should be created with cy_train = np.copy(y_train) rather than cy_train = y_train. In the latter method, the two variables are still entangled and any changes to cy_train will appear in y_train. By explicitly copying over the variables, we "disentangle" the two.

Listing 7-9. Corrupting labels

```
new_values = []
for ind in selected_indices:
    true_label = y_train[ind][0]
    possibilities = [i for i in range(10) if i!=true_label])
    corrupted = np.random.choice(possibilities)
    new_values.append(corrupted)
new_values = np.array(new_values).reshape((len(new_values),1))
cy_train[selected_indices] = new_values
```

In this configuration, 20% of the labels have been corrupted. Generating negative labels for the entire dataset (Listing 7-10) is relatively similar to corrupting the data. We choose a label other than the label currently associated with each training item and use it as the randomly selected *negative label* for that item. Negative labels are stored in the ny_train variable.

Listing 7-10. Generating negative labels

```
ny_train = np.copy(cy_train)
for ind in tqdm(range(len(ny_train))):
    listed_label = cy_train[ind][0]
    possibilities = ([i for i in range(10) if i!=listed_label]
    negative_label = choice(possibilities)
    ny_train[ind] = negative_label
```

The first phase of the NLNL pipeline is to perform negative learning. We will use a standard EfficientNetB3 model with no architecture modification for simplicity (Listing 7-11).

Listing 7-11. Train model on initial negative learning stage. We cannot use ImageNet weights because we're not building a custom top

```
from keras.applications.efficientnet import EfficientNetB3
inp = L.Input((32,32,3))
base_model = EfficientNetB3(
    include_top=True, weights=None,
    input_tensor=inp, classes=10
)
nl_model = keras.models.Model(inputs=inp, outputs=base_model.output)
```

Recall that we need a specialized loss function to train negative learning with. We want to decrease the model's confidence that the training item belongs in the class corresponding to the negative label. If we instead adopt an alternative approach of maximizing the model's confidence that the training item is associated with the class corresponding to the negative label and then perform post-processing negation (e.g., the class with the lowest confidence is the true class), which on the surface seems identical, we run into the problem of forcing the model to learn a fundamentally random phenomenon – which negative label was randomly selected, encouraging overfitting and poor performance.

The general cross-entropy loss equation is, in simple form, $l(y_t, y_p) = -y_t \cdot log\,(y_p)$ (averaged across all $y_t - y_p$ training pairs), where y_t is the true label and y_p is a set of probabilistic predictions. This loss function seeks to maximize the model's confidence of the true class. This correspondingly decreases the confidences of other classes for the same training item, because the softmax output ensures all probabilities sum to 1.

In this case, because we want to *minimize* the model's confidence in the negative label class, we instead use $(y_t, y_p) = -y_t \cdot log\,(1 - y_p)$. The only difference is that the prediction y_p is replaced with $1 - y_p$ such that a higher confidence for the negative label is punished.

Implementing custom loss functions like this in Keras is simple; we define a function that takes in y_true and y_pred and returns the average loss across items using Keras/TensorFlow backend-style functions (Listing 7-12). Because logarithms return NaN when their input is 0, we make sure to use TensorFlow's tf.clip_by_value to ensure that the input is, at its minimum, 1e-5 (or some other arbitrary epsilon value). tf.reduce_mean averages the cross-entropy losses of all the items; passing axis=-1 ensures this. We also make sure to cast y_true and y_pred as float 32 to prevent future type problems.

Listing 7-12. Building a specialized loss function for negative learning

```
import keras.backend as K
def special_loss(y_true, y_pred):
    y_true, y_pred = tf.cast(y_true, tf.float32),
                     tf.cast(y_pred, tf.float32)
    log_inp = tf.clip_by_value(1-y_pred, 1e-5, 1.-1e-5)
    return tf.reduce_mean(-y_true * K.log(log_inp), axis=-1)
```

The model can then be compiled with our special loss and fitted with the negative labels (Listing 7-13). Note that the labels in their current form are not one-hot encoded; we took advantage of this to make label processing when adding noise and generating negative labels more convenient. Normally, we could pass in integer-style labels for multiclass problems and use the "sparse_categorical_crossentropy" loss, which automatically converts the labels to one-hot representation before performing categorical cross-entropy. Because our custom loss function was not built with this functionality, the labels are converted into one-hot form with keras.utils.to_ categorical manually for compatibility with the loss function.

Listing 7-13. Compiling and fitting model for negative learning. We also use a different optimizer – SGD rather than Adam and with particular parameters. This is simply because it seems to perform better. This is a good place to use hyper-optimization. For optimal success, use callbacks and other deep learning good practices

```
import keras.backend as K
from keras.optimizers import SGD
sgd = SGD(learning_rate=0.025, momentum=0.1, nesterov=True)
nl_model.compile(optimizer=sgd loss=special_loss),
nl_model.fit(x_train, keras.utils.to_categorical(ny_train), epochs=100)
```

Note that extensive training is required in each stage for each model version to develop sufficient knowledge representations; insufficient training in one stage will stunt the development of the following stages. This is something to watch out for in all non-end-to-end multi-stage pipeline designs.

The second stage of the pipeline is to perform *selective negative learning* – further negative learning to filter out initial problematic inputs (Listing 7-14). While there are many possible filters to determine which data is problematic or not, for simplicity, we will select training items with which the confidence associated with the positive label is larger than $\frac{1}{10}$. This method begins to filter out corrupted data, which is likely to yield a probability below chance. We can select this subset of data by creating a Boolean mask and applying it to x_train and ny_train to create selnl_x_train and selnl_y_train.

Listing 7-14. Filter out problematic data. For optimal success, use callbacks and other deep learning good practices

```
predictions = nl_model.predict(x_train)
mask = [True if predictions[ind][cy_train[ind][0]] > 0.1 else False for ind
in range(len(x_train))]
selnl_x_train = x_train[mask]
selnl_y_train = ny_train[mask]
```

We can continue to train the model on the updated data (Listing 7-15).

Listing 7-15. Continue to fit data in the selective negative learning phase. For optimal success, use callbacks and other deep learning good practices

```
nl_model.fit(selnl_x_train, selnl_y_train, epochs=100)
```

The third stage of the pipeline is the transition toward *selective positive learning* (Listing 7-16). Selecting data to use in selective positive learning is similar to selecting data for selective negative learning, but we want to use an even more stringent criteria here to ensure that the model is being trained on uncorrupted labels. We will include only data points with which the model assigns at least a 40% confidence to the positive label class. Note that the labels (selpl_y_train) are taken from cy_train – the corrupted positive learning labels, not the negative learning ones.

Listing 7-16. Filter out data for selective positive learning

```
predictions = nl_model.predict(x_train)
mask = [True if predictions[ind][ny_train[ind][0]] > 0.4 else False for ind
in tqdm(range(len(x_train)))]
selpl_x_train = x_train[mask]
selpl_y_train = cy_train[mask]
```

Because we are switching from a negative learning paradigm to a positive learning one, we need to change the loss function such that a higher confidence of the class indicated in the label is rewarded rather than punished. This can be performed by recompiling the model, which still retains its weights (Listing 7-17).

Listing 7-17. Train for the selective positive learning stage. For optimal success, use callbacks and other deep learning good practices

```
nl_model.compile(optimizer='adam', loss='sparse_categorical_crossentropy')
nl_model.fit(selpl_x_train, selpl_y_train, epochs=100)
```

After the SelNLPL pipeline has been completed, this model can be used to separate the dataset into corrupted and uncorrupted datasets (Listing 7-18). Corrupted data is more likely to have a low confidence for the assigned positive label, whereas uncorrupted data is more likely to have a high confidence. We can make this threshold 50% to separate into corrupted and uncorrupted data.

Listing 7-18. Separating corrupted and uncorrupted datasets

```
predictions = nl_model.predict(x_train)

mask = [True if predictions[ind][cy_train[ind][0]] > 0.5 else False for ind
in range(len(x_train))]

clean_x_train = x_train[mask]
clean_y_train = cy_train[mask]
unclean_x_train = x_train[[not boolean for boolean in mask]]
```

Now that a clean dataset has been identified, a new model is trained on the clean labels (Listing 7-19). In this case, it is identical to the model used in the previous SelNLPL pipeline, but it doesn't have to be.

Listing 7-19. Training a model on the clean data to label the unclean data

```
inp = L.Input((32,32,3))
base_model = EfficientNetB3(
    include_top=True, weights=None, input_tensor=inp, classes=10
)
model = keras.models.Model(inputs=inp, outputs=base_model.output)
model.compile(optimizer='adam', loss='sparse_categorical_crossentropy',
              metrics=['accuracy'])
model.fit(clean_x_train, clean_y_train, epochs=100)
```

This model's prediction is used to correct the labels (Listing 7-20). We can extract the integer labels from the model's probabilistic predictions by identifying the index of the maximum probability in each prediction with np.where(). The set of originally unclean, now cleaned, data is fitted into a numpy array and reshaped into the desired shape.

Listing 7-20. Cleaning the unclean data

```
pred_labels = model.predict(unclean_x_train)
unclean_y_train = []
for ind in range(len(unclean_x_train)):
    label = np.where(pred_labels[ind] == np.max(pred_labels[ind]))
    unclean_y_train.append(label)
```

```
unclean_y_train = np.array(unclean_y_train)
desired_shape = (len(unclean_y_train),1))
unclean_y_train = unclean_y_train.reshape(desired_shape)
```

The two datasets can now be concatenated together into a final clean dataset, which can be used to train a final model (Listing 7-21).

Listing 7-21. Concatenating data to form the final cleaned dataset

```
final_clean_x_train = np.concatenate([clean_x_train,unclean_x_train])
final_clean_y_train = np.concatenate([clean_y_train, unclean_y_train])
```

Keeping track of data and models throughout this pipeline can be arduous, but organization is important to prevent mistakes. Mistyping a variable name can lead to a model being trained on the incorrect dataset, which will result in either deceivingly amazing results or frustratingly poor ones.

The NLNL pipeline demonstrates how an initial reframing can be developed into a complete pipeline taking on a difficult problem.

Limited Data: Siamese Networks

Deep learning applications are famously data hungry. A modern neural network often needs to consume and process thousands of data instances per class to obtain a strong representation and understanding of the primary features and attributes associated with that class. Throughout this book, we have explored various methods to decrease the raw amount of new data required for successful training of neural networks. Image augmentation can provide "new data" that the model has not seen before in that exact form. Transfer learning allows for feature extraction skills to be transferred from one general dataset to a more specific dataset, where a smaller dataset can be used to fine-tune/adapt the skills toward a specific context. Self-supervised learning accustoms the model to the fundamental relationships and features of the dataset before it is ever exposed to labels. If you were up to it, you could even train a variational autoencoder to generate new data to increase the dataset size. While these methods are effective in dealing with *small datasets*, they too have trouble with very small datasets.

Few-shot learning is a field of machine learning and deep learning that has recently risen in importance as applications with a high number of classes but few training examples in each class become more important. Think, for instance, of facial recognition models built into applications like face-based logins to phones – the model must be able to verify that your face belongs to you but recognize that your face does not belong to one of the many other possible people that exist. It must do so with very high performance – a model with a 95% accuracy used in login applications is not secure enough – and with very few training examples – it's not a good user experience if the user needs to generate hundreds of training examples!

An even more extreme study within few-shot learning is that of *one-shot learning*. One-shot learning is concerned with classification of inputs given only one instance of each class. For instance, if we are training a model to identify animals under a one-shot regime, the dataset would be comprised of one image from each class (a dog, a cat, a horse, a chicken, etc.) and the model would be expected to generalize all instances of that class from the one image representing the class it was provided. This is, of course, a very difficult task. Advances in one-shot learning and few-shot learning generally, however, have the potential to work toward closing the gap between deep learning and the human mind, which is notably able to generalize concepts and ideas from a much smaller set of instances.

Early work in one-shot learning by Li Fei-Fei et al. in the early 2000s used a variational Bayesian framework relying on the principle that previously learned classes can be used to predict future ones with few examples from the given classes. Later work by Lake et al. in 2013 proposed the *Hierarchical Bayesian Program Learning* method, which can learn drawn figures (i.e., assemblages of lines, curves, and dots) by systematically deconstructing the image and proposing structural explanations for observed pixels. Even through the increasing usage of convolutional neural networks, the most promising approaches to one-shot learning have not used the conventional convolutional neural network architecture but instead have relied upon some level of domain knowledge to fill the knowledge void left by the lack of training data. For instance, if a model is trained to classify MNIST-style digits (10 classes, from 0 to 9), we can build knowledge of the data's structure – line composition, discriminative features, etc. – into the design of the system, like HBPL does.

Thus, an even further challenge on top of one-shot image recognition is to make a step toward *universal one-shot learning*, in which a system can automatically develop knowledge representations from one-shot datasets in all sorts of problem domains (i.e., not needing domain knowledge to be implicitly encoded in the design of the system).

Gregory Koch, Richard Zemel, and Ruslan Salakhutdinov work toward addressing this challenging problem by reframing the problem of one-shot image *classification* to a problem of *similarity* in their 2015 paper "Siamese Neural Networks for One-Shot Image Recognition."[3] Rather than trying to represent each class's features by its own – abstract features floating in a sparse knowledge space – we force the model to learn relationships between the classes, such that knowledge of one class is grounded in knowledge of others. This method allows for a more stable, reliable, and abundant construction of understanding.

Say that you are presented with an apple, an orange, and a grape for the first time. You are told that you must develop knowledge representations of these three entities and apply them when presented to a new object (i.e., recognize if a new object is an apple, an orange, or a grape). You could develop a long list of the features characterizing each object: the apple is round – almost spherical, but slightly oblong – red, with a stem on the top, and a relatively hard exterior. The orange is round – almost spherical, but slightly oblong – orange, with a stub on top, and a soft but firm exterior. The grape is round and small – almost spherical, but slightly oblong – purple, with a small stem on top, and a thin, papery exterior.

There are two key problems with this sort of knowledge representation: firstly, there is evident redundancy in our features that are not relevant for discrimination between the classes. All the objects are round – almost spherical, but slightly oblong, and thus this feature can be discarded because it provides no information for the purposes of classification. Secondly, it relies upon the construction of attributes that are very difficult to come to by themselves. What counts as "soft," for instance, is made meaningful only by the existence of a "hard" object. Likewise, color is meaningful only when contrasted with other colors. Thus, it is impractical to represent knowledge with such few items per class as an identification of various qualities of each class independently.

Rather, we design our method of knowledge representation to capture similarities and differences between each of the items – we are making explicit what is implicit and shoved under the carpet in the previous form of knowledge representation. The orange and the apple are similar in size, whereas the grape is different. The orange, the apple, and the grape are all different in color and in texture. The apple and the grape are similar in the quality of having a stem, whereas the orange has a small stub. These qualities have more meaning because we explicitly represent them as a similarity or difference

---

[3] Gregory Koch, Richard Zemel, and Ruslan Salakhutdinov, "Siamese Neural Networks for One-shot Image Recognition," 2015. Paper link: `www.cs.cmu.edu/~rsalakhu/papers/oneshot1.pdf`.

between objects exemplifying those qualities. Thus, we are able to develop a more stable representation of this knowledge; it is grounded in relationships. When we are presented with a new object, we compare it with each of the previously seen items and assign the object the class of whatever item it is deemed most similar to.

Another perhaps more familiar example: if you have ever been confused by when to use "who" and "whom" and looked it up online, you will likely have found two answers. One usually goes along the lines of "who refers to the subject of a sentence; whom refers to the object of a verb or preposition." The other is "use 'who' when you can substitute it with 'he' or 'she'; use 'whom' if you can substitute it with 'him' or 'her'." The first approach attempts to define the usage of this new concept – who vs. whom – in explicit, non-relational, grammatical terms, using objective pillars of language to support new syntax understanding. The other is relational, taking advantage of other ideas in the contextual space to establish an understanding of this new one. While the former may be more academic and formal, the latter is usually more effective.

In the Siamese network system proposed by Koch et al., a network is trained to classify whether two inputs are the same or not (Figure 7-13). As this is technically a binary classification problem, the network outputs the probability that two inputs belong to the same class.

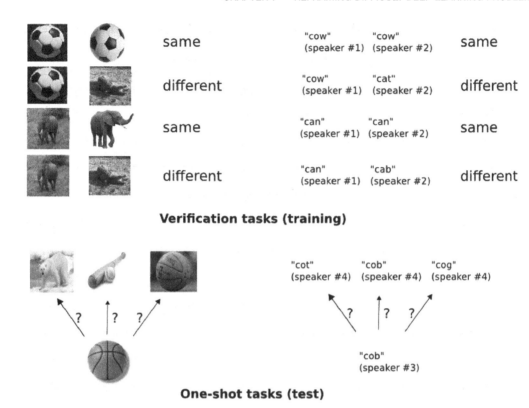

Verification tasks (training)

One-shot tasks (test)

Figure 7-13. *Reframing the classification problem paradigm into one of similarity for one-shot learning. In this case, the network is shown multiple images from the same class; that is, few-shot learning. One-shot learning is generally performed via image augmentation*

The model is trained on these *verification tasks*. Note that because verification tasks are constructed from pairs of data, we can also expand the size of the dataset because one data instance for training the Siamese network is constructed from every pair of data instances. After training, the model predicts the similarity between a test input and every training instance. The class of the training instance with the highest probability of belonging to the same class as the test input is deemed the class of the test input. Note that this paradigm works both in one-shot classification and in few-shot classification.

In order to realize this concept, the *Siamese network* architecture is used (Figure 7-14). Like Siamese twins, the network has two heads to take in the two inputs. The two inputs are processed separately in parallel with one another to extract key features required for comparison. These two parallel branches are entangled by *weight sharing*, in which (recall from Chapter 5, Efficient Neural Architecture Search case study) weights across different parts of the network are the same and updated identically. This weight sharing ensures that the order of input does not matter and that both inputs are processed in the same manner. After the inputs have been processed in parallel, their encoded features are compared via a distance layer, which computes the distance between the encoded representation. This distance is then further processed via an output layer to yield the final probability/similarity.

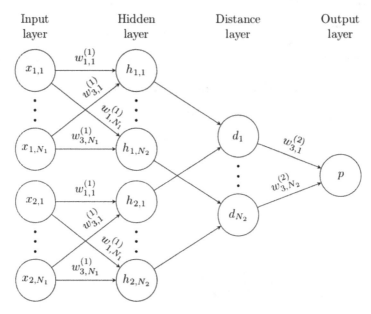

Figure 7-14. *Simplified Siamese network architecture*

The Siamese network design used by Koch et al. uses the standard sequence of alternating convolution and pooling operations (Figure 7-15).

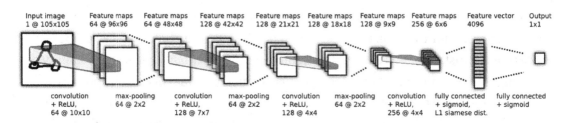

Figure 7-15. *Convolutional branch of the Siamese network used in the original paper by Koch et al.*

The Siamese network system performs well in multiple contexts. Consider, for instance, the Omniglot dataset (Figure 7-16), which contains images of characters in various alphabets – Aurek-Besh, Greek, Hebrew, Korean, Latin, Malay, etc. The convolutional Siamese network performs better than almost all other algorithms and models in one-shot learning on the Omniglot dataset (Table 7-3).

Figure 7-16. *Sample of 20 classes from the Omniglot dataset*

Table 7-3. *Comparison of Siamese network method against other one-shot learning methods*

Method	One-Shot Test Acc.
Humans	95.5%
Hierarchical Bayesian Program Learning	95.2%
Affine model	81.8%
Hierarchical deep	65.2%
Deep Boltzmann machine	62.0%
Simple stroke	35.2%
1-nearest neighbor	21.7%
Siamese network	92.0%

As noted prior, the Siamese network performs slightly poorer than the Hierarchical Bayesian Program Learning, with the advantage of being applicable to all sorts of other datasets and contexts.

Let's begin by constructing a dataset for the Siamese network from the MNIST dataset (Listing 7-22). We will begin by sampling a small quantity of items from each of the classes – in this case, ten training instances for each of the ten-digit classes. We do this first by selecting all indices with a *y* value that match a digit label and then use np.random.choice to randomly select a certain number of indices from the index list.

Listing 7-22. Loading and sampling MNIST data

```
# configurations
class_size = 10

# load MNIST data
(X_train, y_train), (X_test, y_test) = keras.datasets.mnist.load_data()

# select x instances from each class
samp_x_train, samp_y_train = [], []
for digit in range(10):
    indices = (y_train == digit).nonzero()[0]
    selected = np.random.choice(indices, size=class_size)
    samp_x_train.append(X_train[selected])
    samp_y_train.append(y_train[selected])
```

We need to make some modifications to the dataset. Firstly, we convert both lists of data into numpy arrays and scale the data from between 0 and 255 to between 0 and 1. Afterward, we reshape the data into the appropriate form such that there are *number classes* × *class size* data instances and that images have one channel (Listing 7-23).

Listing 7-23. Scaling and reshaping data

```
samp_x_train = np.array(samp_x_train)/255
samp_x_train = samp_x_train.reshape((10*class_size, 28, 28, 1))
samp_y_train = np.array(samp_y_train)/255
samp_y_train = samp_y_train.reshape((10*class_size, 1))
```

We loop through all unique combinations of indices to other indices (Listing 7-24); the second index is selected from all indexes existing *after* the first sampled index to prevent repeating combinations twice. If the labels of the two indexed data points are equal, we use the label 1 to indicate that the two images belong to the same class. Otherwise, the label is 0.

Listing 7-24. Generating Siamese network-style pairs

```
# generate pairs
fs_x_train_1, fs_x_train_2, fs_y_train = [], [], []
indices = list(range(10*class_size))
for ind_1 in indices:
    label1 = samp_y_train[ind_1]
    for ind_2 in indices[ind_1:]:
        label2 = samp_y_train[ind_2]

        # append x
        fs_x_train_1.append(samp_x_train[ind_1])
        fs_x_train_2.append(samp_x_train[ind_2])

        # append similarity label
        if label1 == label2:
            fs_y_train.append(1)
        else:
            fs_y_train.append(0)
```

We can convert the three sets of data into numpy arrays for usage in the Siamese network (Listing 7-25).

Listing 7-25. Converting to numpy arrays

```
fs_x_train_1 = np.array(fs_x_train_1)
fs_x_train_2 = np.array(fs_x_train_2)
fs_y_train = np.array(fs_y_train)
```

We will be using compartmentalized design to build the Siamese network (Listing 7-26). The two parallel branches will be built as separate sub-models that are arranged together into a larger Siamese network. The purpose of this is to implement weight sharing, which we will discuss later. Each parallel branch will be a standard alternation between convolution and pooling layers. After convolutional processing, the data is flattened and mapped to a 16-dimensional encoded representation. The branch is arranged into a model that takes in an input image and maps it to a vector encoded representation, which can later be used in addition to another vector encoded representation to calculate the distance between encoded representations.

Listing 7-26. Function to create a parallel branch model

```
def parallel_branch():
    inp_layer = L.Input((28,28,1))
    x = L.BatchNormalization()(inp_layer)
    x = L.Conv2D(32, (3,3), activation='relu')(x)
    x = L.Conv2D(32, (3,3), activation='relu')(x)
    x = L.MaxPooling2D((2,2))(x)
    x = L.Conv2D(64, (3,3), activation='relu')(x)
    x = L.Conv2D(64, (3,3), activation='relu')(x)
    x = L.MaxPooling2D((2,2))(x)
    x = L.Flatten()(x)
    x = L.Dense(16, activation='relu')(x)
    branch = keras.models.Model(inputs=inp_layer, outputs=x)
    return branch
```

Next, we'll implement the distance layer (Listing 7-27), which does not exist already in Keras/TensorFlow. Luckily, we can take advantage of the keras.layers.Lambda function, which allows us to define a function using backend functions as a layer. The distance function takes in the two encoded vector representations and uses Keras backend functions to return the L2 Euclidean distance between the two, defined as $\sqrt{\Sigma\left(e_{1i}+e_{2i}\right)^2}$ for the two encoded representations e_1 and e_2. Because of complications with calculating the gradient of $f(x)=\sqrt{x}$ at $x = 0$ (i.e., when two images are the same and $e_1 = e_2$), we set $x = max\,(s, \epsilon)$ before square rooting, where s is the sum of squared differences and ϵ is a very small nonzero number. If we do not take this measure, the loss will become NaN the first instance that the gradient is calculated when both inputs are the same and thus the encoded representations are identical.

Listing 7-27. Function to calculate distance between representations

```
import keras.backend as K
def distance(representations):
    reps1, reps2 = representations
    squared = K.sum(K.square(reps1-reps2), axis=1, keepdims=True)
    return K.sqrt(K.maximum(squared,K.epsilon()))
```

We can construct the Siamese network relatively simply using these compartments and functions (Listing 7-28). To build multi-input models, we construct two Input layers and list the inputs in a list when aggregating layers into a model. To institute weight sharing, we first begin by instantiating one model, which is used to process each of the inputs. We use the same instantiated model to process both inputs, thus implementing weight sharing. The resulting representations are passed into the distance function, the output of which is further processed into an output.

Listing 7-28. Creating Siamese network with weight sharing

```
inp1 = L.Input((28,28,1), name='inp1')
inp2 = L.Input((28,28,1), name='inp2')
branch = parallel_branch()
reps1 = branch(inp1)
reps2 = branch(inp2)
dist = L.Lambda(distance)([reps1, reps2])
out = L.Dense(1, activation='sigmoid')(dist)
model = keras.models.Model(inputs=[inp1, inp2], outputs=out)
```

Plotting the Siamese network architecture visualizes how Keras interprets our weight sharing implementation (Figure 7-17).

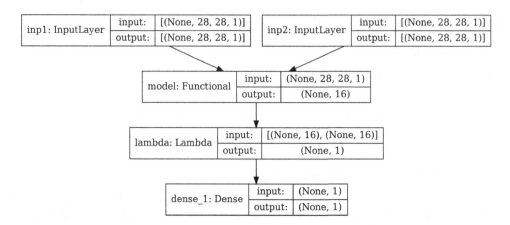

Figure 7-17. *Keras diagram of Siamese network architecture*

To train the model, we compile and fit the model (Listing 7-29). To indicate which data corresponds to which input branch, we pass in a dictionary where the key is the name of the input layer and the value is the corresponding array.

Listing 7-29. Compiling and fitting the Siamese network

```
model.compile(optimizer='adam', loss='binary_crossentropy',
              metrics=['accuracy'])
model.fit({'inp1':fs_x_train_1, 'inp2':fs_x_train_2}, fs_y_train, epochs=100)
```

Dozens of epochs of training yields a near-perfect Siamese network that can be used to perform classification tasks in few- and one-shot learning contexts. Siamese networks also have applications in similarity problems, like text or image matching.

The design of the Siamese network demonstrates how problems can be elegantly reframed to adjust to shortcomings of existing methods through relatively simple adaptations. Moreover, it highlights yet again the experimental and intellectual freedom with which one must approach difficult problems to successfully reframe them.

Key Points and Epilogue

In this chapter, we dove into creative problem-solving with deep learning by reframing the problem with three examples: DeepInsight, which reframes how traditional structured data is represented as images; the Negative Learning for Noisy Labels (NLNL) pipeline, which reframes the positive learning paradigm as negative learning; and the

Siamese network, which reframes classification as a similarity problem. The primary goal of this chapter was to emphasize the key role data plays in *using* deep learning in real applications. Generally, it is data that dictates how the deep learning system is developed, not vice versa – but we can use problem-solving skills and the versatility embedded within deep learning methods to reframe current methods to adapt to the circumstances of the data.

The key theme unifying the successful case studies covered in this chapter is the spirit of *reframing*. The first step to reframing is to identify an unspoken boundary that exists as an obstacle but has not been challenged, either because no one has tried to or because at first appearance it may not seem like it can or should be challenged. Here, while you are reading these words, you are encouraged to ponder seriously about unspoken, implicit boundaries within the massive, ever-expanding realm of deep learning – without thinking too much about implementation, for that is a task for later – and the ideas of reframing will blossom.

This chapter, along with this book, concludes in these last few pages. Throughout this book, we have covered a wide swath of methods and techniques currently at the forefront of deep learning work. Ultimately, these are ideas and tools for you not only to use but also to combine, break, piece together, experiment, and innovate with. If you push enough at an implicit boundary, chances are you'll find something meaningful in reframing it. Deep learning is still on a meteoric rise, and it will be propelled not by those that are only users of existing discoveries but by those that build upon previous advancements to pioneer new ideas.

Index

A

Acquisition function, 267

.add() method, 33

Adjusting learning rate, 22

Adversarial learning, 95

alter_data function, 93

Altered task datasets, 61, 94

Anchor points/layers, 333, 342

Articulation of valued features, 163

Asymmetric convolutions, 374

Asymmetric factorization module, 374

Attack performance metrics, 101

Autoencoders *See also* Variational
    autoencoder
  components, 118, 119
  data forms, 145
  denoising (*see* Denoising
    autoencoders (DAE))
  dimensionality reduction
    implementation, 164
    intuition, 161–164
  encoding component, 122
  feature generation, 164–172
  image data
    convolutional autoencoder vector
    bottleneck design, 137–139
    convolutions without
    pooling, 133–136
    convolutions with pooling and
    padding, 140–145
    "enlarging" operation, 127

"reducing" operation, 127
  shape structure and
    transformations, 127–133
  implementation, 121
  input and output, 119
  intuition and theory, 116–120
  pretraining
    implementation, 160
    intuition, 156–160
    TabNet (case study), 189–193
  tabular data, 121–127

Auto-Keras
  API, 300
  architecture of model search, 309
  blocks, 306
  complexity principle, 297
  edit distance neural network
    kernel, 299
  head/output block, 303
  input node, 302
  NAS system, 298–302
  NAS with custom search space,
    305–309
  NAS with nonlinear
    topology, 309, 311
  processing block, 302
  simple NAS, 302–305
  SMBO, 298
  user-specified epochs, 300

Automatic image dataset, 44, 45

Automation, 329

Printed in the United States
by Baker & Taylor Publisher Services